The Crab Nebula

By the same author

The Crab Nebula

SIMON MITTON

CHARLES SCRIBNER'S SONS
New York

Library of Congress Cataloging in Publication Data

Mitton, Simon, 1946–
  The Crab Nebula.

  Bibliography: p.
  Includes indexes.
  1. Crab Nebula.   I. Title.
QB855.M57   523.1'12   78-23381
ISBN 0-684-16077-3

First U.S. edition, 1979
1 3 5 7 9 11 13 15 17 19 I/C 20 18 16 14 12 10 8 6 4 2

Printed in Great Britain.

Contents

Illustrations

Author's Preface

Of all the objects beyond our solar system the Crab Nebula is the most fascinating to modern astrophysics. Within the tangled web of this unique object there lies a spinning pulsar, the densest known body. Thirty times a second it rotates, flashing the message of its existence to the universe.

The Crab Nebula is a cosmic battleground. Forces of physics oppose each other according to the laws of nature. In the pulsar enormous gravitational pressure and strange quantum forces are equally matched. In the nebula electromagnetic effects contribute to a beautiful celestial light show. The Crab Nebula is counted among the brightest objects in the sky at radio and X-ray frequencies, and consequently it has received much attention from professional astronomers.

In many respects the Crab Nebula is a cosmic laboratory, and it is this theme that unifies the present book. Processes that cannot be simulated in an ordinary laboratory take place out there, 6500 light years away. Magnetic and gravitational forces far beyond anything that experimental physicists will ever achieve on Earth control the nebula and pulsar.

My aim in this book is to give a biography of the Crab Nebula as told in the several hundred research papers and monographs devoted to it in particular and supernovae in general. The approach is entirely non-mathematical because I want this exciting story to be accessible to a wide audience of amateur astronomers and students. The glossary is to help those readers unfamiliar with certain terms, while the bibliography will enable others to explore the research I used.

The book was mainly written between January and May 1977, but it includes some material as recent as October 1977.

November 1977

Simon Mitton
Institute of Astronomy
University of Cambridge
England

Acknowledgements

Many colleagues assisted during the preparation of this book, contributing ideas, criticism, and encouragement. The suggestion I should write the book in the first instance came from Dr T. E. Faber of the Cavendish Laboratory Cambridge, back in 1971. However, pressure of other projects forced a postponement until late 1976. I am grateful to the London publishers, Faber and Faber, for agreeing to the project after so long a delay on my part.

Among the colleagues who aided me I should like to thank Dr John Whelan (Institute of Astronomy, Cambridge), Dr Gregory Benford (University of California), Dr Kris Davison (University of Minnesota), Dr Richard Stephenson (University of Newcastle-upon-Tyne), Professor Antony Hewish (Mullard Radio Astronomy Observatory) and Professor Franz Kahn (University of Manchester) for critically reading parts of the manuscript. Illuminating conversations with Dr Virginia Trimble (University of California) and Professor F. Graham Smith (Royal Greenwich Observatory) are acknowledged. Professor T. Gold (Cornell University) and Dr Franco Pacini (European Southern Observatory) kindly supplied their reminiscences of early theoretical developments. A bibliography assembled over many years by Dr John Shakeshaft (Mullard Radio Astronomy Observatory) greatly aided the literature search. Several individuals and observatories contributed material for illustrations and they are acknowledged in the captions. It is a pleasure to thank the Hale Observatories and the Mullard Radio Astronomy Observatory for the considerable trouble they went to in supplying illustrations. I must add that all blemishes remaining, despite the help of the above, in the final book are my sole responsibility.

The final typescript was beautifully prepared by Sally Roberts.

Finally I thank the Institute of Astronomy and my family; without their help this project could not have been undertaken.

1 A.D. 1054: A star explodes

1.1 Oriental watchers of the sky

On a fine summer night in Imperial China in the year 1054 the court astrologers diligently observe the heavens. Constellations that we know as Ursa Major, Perseus, Andromeda and Pegasus sparkle in the clear air. The Milky Way crosses the sky above. About two hours after midnight attention is focused on Taurus: there, beckoning between the horns of the constellation that we call the Bull, a new star blazes forth, outshining all other stars and even the planets Jupiter and Venus. The astrologers are in a state of excitement and confusion. What can this appearance of a brilliant 'guest star' mean, and what are its implications for the Emperor and the Sung dynasty?

Nearly two months later, on August 27 the distinguished astrologer Yang Wei-te reports thus to the anxious Emperor:

> 'Prostrating myself before your Majesty I hereby report that a guest star has appeared; above the star in question there is a faint glow, yellow in colour. If one carefully examines the prognostications concerning the emperor, the interpretation is as follows: The fact that the guest star does not trespass against *Pi* the lunar mansion in Taurus and its brightness is full means that there is a person of great wisdom and virtue in the country. I beg that this be handed over to the Bureau of Historiography.'

The attendant officials present their congratulations to the great astrologer; the Emperor is pleased and orders that the records be dispatched to the official chroniclers in the Bureau of Historiography.

This particular account of the appearance of a new star in the heavens is preserved for posterity in 'The Essentials of Sung History' compiled by Chang Te-hsiang. From the text it seems possible that astrologer Yang Wei-te was a sycophant using the unexpected event for his own political ends. His mention of the star's yellow colour, for example, cannot be relied upon because yellow was the imperial colour of the Sung dynasty. Apart from the divinations recounted above, the chronicle also gives valuable astronomical details. The Director of the Astronomical Bureau said of the guest star 'It was visible in the daytime, like Venus. It had pointed rays on all sides and its colour was reddish-white. Altogether it was visible for 23 days (in daylight).'

Unlike the sky-gazers of Europe, astronomers in the Far East— China, Japan, and Korea—routinely noted down transient celestial phenomena for hundreds of years. Their observations survive to the present day in a number of sources, most of them astronomical treatises of the dynastic histories. There are Chinese records that go back to 200 B.C., and even earlier, whereas the Japanese and Korean documents commence in about A.D. 800; consequently there is frequent duplication of the data after A.D. 800, a fact that helps historians of astronomy to interpret these fascinating records today. Many of the astronomical observations concern the appearance of new stars in the sky, the word star being used loosely to include comets as well as stellar objects. We can appreciate that any change in the patterns of the stars would have been of great importance to the professional astrologers employed by the emperors. Even these astrologers, however, can scarcely have imagined the great impact that their observations of the new star in A.D. 1054 would have on the course of science nine centuries later.

The star of A.D. 1054 was not the brightest to burst on the scene in historic times: more brilliant jewels shone out in A.D. 185 and 1006. Perhaps a few astrologers were fortunate enough to witness both events of A.D. 1006 and 1054. Also some of the cometary apparitions, described as 'broom stars' or 'sweeping stars' (*hue-hsing*), were bright enough to remain visible in daylight; this is definitely true for certain of the returns of Halley's Comet, for example. In the Chinese chronicles the expression 'guest star', or 'visiting star' (*k'o-hsing*) was generally, but not exclusively, used to indicate a new star-like object. The term is frequently synonymous with what modern astronomers call a nova or supernova. Very careful searches of all the extant records from the Far East have shown that about 75 authentic guest stars were recorded between 532 B.C. and A.D. 1064. This total excludes the sightings that can be reliably attributed to comets, or that are of a doubtful nature.

Altogether there are eight historical sources of the guest star of A.D. 1054; five of these are independent accounts, four from China and one from Japan. It is interesting to see how these vary in their descriptions of the object and its behaviour. In the translated extracts given here explanatory material that is not part of the original texts is indicated by enclosing it in brackets.

The astronomical treatise in the annals of the Sung dynasty, compiled around A.D. 1345, contains the following references:

> In the first year of the period Chih-ho (A.D. 1054), the fifth moon, the day of *chi-ch'ou* (July 4) (a guest star) appeared approximately several inches (roughly 0.3 to 0.5 degrees) to the south-east of *T'ien-kuan* (the star Zeta Tauri). After a year or more it gradually vanished.

An identical account appears in an encyclopaedia written about A.D. 1280, and it is apparent that the authors of this summary had access to the sources available to the official historians. Another passage in the Sung dynasty's history enables us to fix the precise date (1056 April 17) at which the guest star finally vanished from human vision:

> On the day *hsin-wei*, first year of the Chia-wu reign, third month, the Director of the Astronomical Bureau reported that from the fifth moon of the first year of the period Chih-ho (i.e. 1054 July 4) a guest star had appeared in the morning in the eastern heavens— which had only now become invisible.

From these two records we deduce that the star lasted 653 days before fading from naked eye visibility in the night sky.

A record that predates all others is found in a Sung chronicle due to one Li Tao, who died in A.D. 1184. He states that a guest star flared up 'to the south-east of *T'ien-kuan* (Zeta Tauri), possibly several inches away'. However, Li Tao's account is so similar to that of the official annals that they cannot be considered as independent sources. Nevertheless, the story of the astrologer Yang Wei-Te quoted above is in an unrelated chronicle and gives confirmation of the event.

The next mention, in chronological order, is in the memoirs of the Liao kingdom written around A.D. 1350. This semi-nomadic kingdom existed in the extreme north of China from A.D. 937 to 1125. On May 10 in A.D. 1054 there was a total eclipse over central China which would have given an impressive partial eclipse further north, and this is linked to the sudden appearance of a guest star and the subsequent death of their king in the following passage:

> (1055 August 28) the king died. Previously there was a solar eclipse at midday and a guest star appeared in the (Pleiades). The deputy officer in the Bureau of Historiography said: This is an omen that (the king) will die. The prediction indeed came true.

The date of the star fits reasonably with the Sung information but the location disagrees by 20°. However the historian is mainly concerned with the death of the king, rather than astronomical observations, and may have simplified the account by mentioning the Pleiades, one of the few well-known star groups. The oriental sources refer to no other guest star around this time, so in view of the ample evidence that one appeared near to Zeta Tauri the discord is not serious.

We find further confirmation of the Chinese observations in Japanese sources. For example, around A.D. 1235 the poet–courtier Sadiae Fujiwara wrote:

> Second year of the Tenki reign period, fourth month, after the middle decade. At the hour *ch'ou* (i.e. 1—3 am) a guest star appeared in the degrees of *Tsue* and *Shen*. It was seen in the east and flared up at *T'ien-kuan*. It was as large as Jupiter.

The star Zeta Tauri (*T'ien-kuan*) lies in 'the degrees of *Tsui* and *Shen*'. The phrase 'as large as Jupiter' really means that the guest star rivalled the brightness of Jupiter. Persons with no astrophysical knowledge frequently call bright stars 'large stars'. One puzzle in this Japanese observation is that the date corresponds to May 29, not early July, and yet it is quite inconceivable that the Japanese would have sighted it five weeks before the Chinese astronomers. In any case Zeta Tauri would not have been visible on May 29 as it was then too close to the sun. The only solution is to assume that the text is corrupt: it is not unusual for errors of a month to occur in these oriental astronomical treatises.

Another Japanese chronicle, of unknown date and authorship recounts precisely the same story as courtier Sadiae Fujiwara and is presumably from the same original source. The second possibly independent Japanese version is found in a massive history finished in 1715; however, this is such a late work that we cannot be sure that the information was not copied from the Japanese documents.

1.2 Interpretation of the historical records

From the five sources that are probably independent we can now try to assemble a reasonably consistent picture of the stellar outburst of A.D. 1054. This is not a trivial exercise because there are discrepancies in all the stories.

In establishing the identity of the guest star it is crucially important that the four records mention that the new star was near to *T'ien-kuan*. This is an object that sinologists have shown conclusively is the star Zeta Tauri. For example, 'The Secret Garden of the Observatory', an astronomical text originally compiled in A.D. 580 and revised in A.D. 1050, gives categorical proof that this identification is correct. (Zeta Tauri is the more southerly and fainter member of the pair of stars that form the tips of the 'horns' of the constellation figure the Bull.) Now we have to deduce the position of the guest star relative to Zeta Tauri. All accounts emphasize the proximity to Zeta Tauri, but the A.D. 1345 version, in the Sung annals, states explicitly that the separation was 'several inches'. This use of a linear unit to define an angle may surprise us, but in the astronomy of the Far East such a method was regularly used. By checking certain observations made by Chinese astronomers (e.g. the measured distances of certain stars from the ecliptic, and planetary conjunctions) certain scholars have concluded that 'an inch' corresponds to about 0.1 degrees on the sky. Furthermore 'several' generally means 3, 4 or 5 in the writings of the time, so we conclude that the new star was some 0.3–0.5 degrees from Zeta Tauri.

What can we learn of the nature of this star? Can we be certain, for example, that the astrologers were not seeing a comet rather than a true star? The texts all avoid using the usual expression for a comet, so if it was in reality such a body it cannot have had a discernible cometary tail. Although the phrase 'pointed rays' is used once, we can safely assume that this is a correct description of what a bright starry object looks like to the human eye. We can commonly notice this effect with Venus near to its maximum brightness. What is especially important is the lack of any mention of motion relative to the stars during the twenty-one months of visibility. We are certain, from the account of Yang Wei-te, that 'the star does not trespass against the lunar mansion in Taurus', that the object was more or less stationary for at least two months after its appearance. Static behaviour for so long is quite inconsistent with the sudden appearance of a brilliant daylight comet, which may travel several degrees in a single day. Furthermore, daylight comets, such as the one in 1910 and Comet West of 1975, possess enormous tails that are spectacular sights; obviously the oriental scribes were not writing about a daylight comet.

Several lines of evidence demonstrate conclusively that the A.D. 1054 event was not a comet. This inevitably means that it must have been due to an exploding star: either a nova or a supernova. We shall look at the nova hypothesis first, from the viewpoint of modern astrophysics.

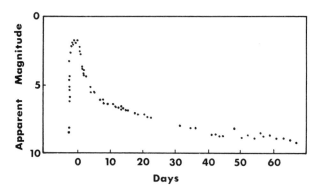

Fig. 1 Light curve of Nova Cygni 1975, compiled from observations from several sources. This particular nova was characterized by the very large range of its outburst, which exceeded 19 magnitudes.

Nova outbursts probably occur exclusively in binary star systems. They are triggered when one star of the pair reaches a point in its life cycle at which it starts to swell up and thus to transfer matter to the companion star. The unexpected influx of material to the companion star upsets its stability. Under certain conditions an explosion results. An important feature of the nova phenomenon is that the brightest novae show the fastest decline from maximum intensity. The faster they rise the harder they fall. Nova Cygni, which flared in 1975 to about second magnitude, and which underwent a total change of at least 19 magnitudes, dramatically declined in only a few days. The A.D. 1054 star was a *daylight* spectacle for twenty-three days, which implies a magnitude of −4 or slightly brighter at the end of the daylight phase: we cannot be sure how bright it was on 1054 July 4, but −4 is the lower limit with −5 more likely. A nova behaving in this fashion would have to be within sixty light years of Earth: if it were any further away it could not have exceeded the threshold of daylight detectability for three weeks. On the average we would only expect a nova this close to Earth once every 30,000 years, so the statistical evidence for a nova hypothesis is somewhat negative. Furthermore, if the star were a nova it would have been as bright as the Moon initially, and we would expect to find very dramatic accounts in the historical records.

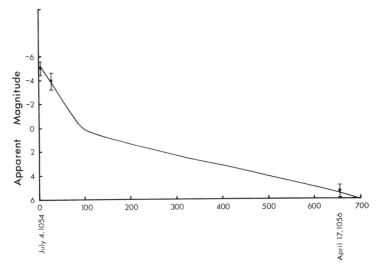

Fig. 2 Light curve of the A.D. 1054 explosion as deduced from contemporary records. The explosion in this case showed a much slower decline than is typical for nova explosions that reach naked-eye visibility.

Astronomers have carefully searched the vicinity of Zeta Tauri for any remaining evidence of a nova in the last thousand years. What they have looked for are binaries containing a highly-evolved red star and a hot white dwarf, because nova explosions are the result of a red star's extended atmosphere crashing into the gravitational prison of a tiny white dwarf. No such binaries exist near Zeta Tauri. Another approach is to look for stars with discernible motion relative to Zeta Tauri. The apparently changeless heavens are in fact in ceaseless motion. A star within sixty light years of our Sun would certainly have a measurable motion relative to other stars. The only three stars near Zeta Tauri with large enough proper motions cannot possibly have engaged in pyrotechnic displays at any time in their history let alone the last millennium. This is because they are normal stable stars, and one of them is actually very like our Sun.

1.3 Identification with the Crab Nebula

We can safely conclude that the star of A.D.1054 was neither a comet nor a nova, and by default we must accept it was a supernova. Fortunately there is strong evidence in support of this conclusion.

Supernova explosions occur in solitary stars that are much more massive than our Sun. Towards the end of its life the star is unable to produce enough energy to keep its interior sufficiently hot. The declining energy production causes the internal temperature and pressure to fall. Soon the object cannot support its own weight, and so its core begins to collapse under the force of gravity. The implosion of the core sets up mighty nuclear explosions in the outer layers of the star. Travelling at thousands of kilometres per second, the stellar atmosphere is ejected. For a short time the luminous energy from the shattered wreck of the dying star is equivalent to the output of millions of normal stars. Thus can a new star shine forth from afar.

In a supernova explosion the climb to maximum brilliance is somewhat slower than for a nova. This more leisurely ascent to the dramatic climax increases the chance that the object will be discovered a few days before it reaches peak brightness. A supernova about 7,000 light years from Earth would have a maximum apparent brightness of—5 magnitudes and be visible by day for three weeks, behaviour that is entirely in accordance with the oriental sources. Supernova explosions thrust enormous quantities of gas into space, and emission from this remains detectable to optical and radio telescopes for thousands of years. Hence the event of A.D. 1054 must have left a remnant if it were a supernova explosion.

The remarkable Crab Nebula, which lies to the north-west of Zeta Tauri, has long been accepted as a small supernova remnant. Edwin Hubble, better known in astronomy for his epochal contributions to the study of galaxies beyond the Milky Way, first suggested in 1928 the association of this relic with the events of A.D.1054. Prior to this, in 1921, J. C. Duncan had reported the expansion of the Crab Nebula. Hubble noted that at the observed rate of expansion 900 years would suffice to allow the nebula to reach its present size.

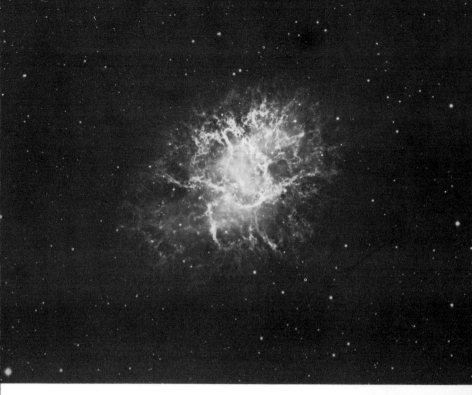

1. The Crab Nebula, showing the tangle of filaments
(Courtesy of the Royal Astronomical Society)

In the 1940s the additional oriental evidence reviewed above was marshalled to support the association of the guest star with the birth of the Crab Nebula. Although this idea came to be generally accepted, discrepancies exist. Perhaps the most serious, and certainly the main focus of attacks on the hypothesis, is the fact that the Crab Nebula is 1.1 degrees north-west of Zeta Tauri, whereas the Sung dynastic record indisputably states that the guest star was south-east of Zeta Tauri. There are at least two ways to wriggle out of this difficulty. One possibility arises because the contents of the annals of the Sung Dynasty are said to have been corrupted by the Bureau of Historiography. Alternatively, the astrologer responsible for writing up the report may have transposed the relative positions of the two stars. Observational astronomers, both professional and amateur, will know perhaps from personal experience that the latter can and does happen. It has certainly occurred elsewhere in astronomical treatises from the orient. Since independent accounts demonstrate the nearness of the guest star to Zeta Tauri, and hence to the Crab Nebula, we need not be unduly worried.

Despite all the evidence in favour of the identity of the genesis of the Crab Nebula with the guest star, some scholars have continued to challenge the consensus view. Mainly they have argued that not a single description in the ancient literature fits the Crab Nebula in all respects. In this objection they are formally correct. However the totality of the reports certainly fits if pieces of information without independent confirmation are rejected. Acceptance of the alternative brings grave problems of a different kind. We must explain why the brilliant star that did generate the Crab Nebula is nowhere recorded, because the work of E. J. Duncan showed that it must have happened within a century of A.D. 1054. We must also account for the fact that sensitive radio surveys have failed to detect any explosion remnants near to Zeta Tauri apart from the Crab Nebula. Moreover, rejection of the Crab Nebula–A.D. 1054 association implies that two supernovae occurred in quick succession in the same part of the sky, which is an incredibly unlikely coincidence. It is easier to believe that all those centuries ago astrologers, civil servants, and scribes made minor errors than to hold the view that the guest star of A.D. 1054 did not mark the birth of the Crab Nebula.

1.4 European and Arabian observations

Because of the obvious importance of the A.D. 1054 guest star to oriental observers, the lack of any mention whatsoever of strange celestial phenomena in European and Arabic records is at first sight extraordinary. It seems amazing that a transient star visible for three weeks by day and almost two years by night should not have received a mention anywhere. It is ridiculous to suppose that observing conditions could have been uniformly bad over the whole of Europe and the Near East for so long a span of time, so we can be certain the Europeans and Arabs did see the new star. Why did they not record the event?

The absence of any information in western sources has to be considered against the prevailing view of science. Europe was languishing in the intellectually stifling atmosphere of the Dark Ages, still four centuries away from the Renaissance and hopelessly constrained by Aristotelian philosophy. In the Aristotelian doctrine the starry realms were regarded as essentially changeless, with the result that little scientific interest was taken in transient astronomical events, even by educated persons in Europe and the Arab lands. It is true that many records of comets have survived, but these denizens of the solar system were incorrectly regarded as a feature in Earth's atmosphere. Hence they were observed with a mixture of interest and suspicion; Halley's comet of A.D. 1066, for example, is included as a portent in the Bayeux tapestry. Faith in the idea that Europeans and Arabs had no interest in the guest stars is reinforced by the fact that very few of those seen in the orient are recorded in western chronicles, although comets are frequently mentioned. This further strengthens the view that the A.D. 1054 outburst was definitely not a bright comet. In Arab manuscripts only the supernova of A.D. 1006, which was brighter than the one forty-eight years later, can be found. Certain Russian chronicles have data on transient events such as eclipses and sunspots, but it is thought by some that any accounts relevant to the birth of the Crab Nebula were destroyed when the Mongols sacked Kiev. Finally keep in mind the fact that astronomers in the orient operated under royal patronage, frequently as employees of the Emperor. They, therefore, had a much stronger incentive to record their observations; indeed there was even a civil service to relieve them of that irksome chore! Europe had to wait some 600 years for its first Astronomer Royal, and no supernova has blazed forth in the Milky Way since that time.

1.5 The Arizona rock paintings

A millennium ago native astronomers were active in the Americas, and even now the Navajo Indians in northern Arizona practise astronomy in a way that has probably remained unchanged for centuries. In the spectacular Canyon de Chelley in north-eastern Arizona, for example, there are elaborate paintings of the starry sky on cave ceilings and rock overhangs. Most of these are less than three hundred years old but they must have their origin in an older tradition. The Navajo are also noted for their beautiful sand paintings that incorporate celestial symbols. Among the unmistakable representations on Navajo gourd rattles and in Hopi art are the Sun and Venus, crescent moon, Orion, Pleiades and the Milky Way. Primitive societies tend to be ultra-conservative in their rituals so it is not unreasonable to suppose that the American Indians of a thousand years ago were inclined to record astronomical phenomena, although no unbroken written or oral thread connects our culture and time with theirs.

In 1955 William C. Miller astonished everyone who took an interest in the Crab Nebula with an announcement that he had found two rock paintings in northern Arizona that appeared to be drawings of the outburst. Each petroglyph (rock picture) discovered by Miller consists of a crescent and a circle, which Miller interpreted as the crescent moon and nearby supernova. This deduction is superficially plausible because on the morning of 1054 July 5, the crescent moon, as seen from western North America, was remarkably close to the supernova. The hypothesis is reinforced by the fact that crescents are rarely found in meso-American rock art. The date when the Arizona paintings were made is not known, but pottery fragments indicate that the general area was inhabited for several centuries, including the year A.D. 1054. What is especially interesting about these Arizonan pictures is that the Moon moves through its own diameter relative to the distant stars in only one hour. In astronomical terms this is a rapid relative motion. Accurate computations have shown that a conjunction between the crescent Moon and the brilliant supernova could only have been seen in the western parts of the American continent. Surely such a dramatic juxtaposition of these two objects must have made a great impression on those primitive watchers of the skies.

Miller's speculations encouraged archaeologists and astronomers to search for more renderings of the Crab Nebula outburst in Indian art, and several further pictures were located in the early 1970s.

Among these are the handiwork at Fern Cave in the Lava Beds National Monument of northern California, which Man has inhabited since 1500 B.C. There he has left an impressive display of his art on the walls and ceiling of the cave. One rock panel has a charcoal sketch of a crescent moon with horns pointing towards a circle. Eight kilometres south-west of Fern Cave is Symbol Bridge where a pictograph contains three crescents adjacent to round objects, and some researchers have claimed that this is reminiscent of the supernova–crescent moon conjunction.

In June 1972 archaeologists from the University of New Mexico found a beautifully executed painting in red hematite, of a handprint, inverted crescent moon, the sun, and a bright star at Chaco Canyon (north-western New Mexico). Studies of the growth rings of timber in the neighbourhood have shown that this sacred site of the Pueblo

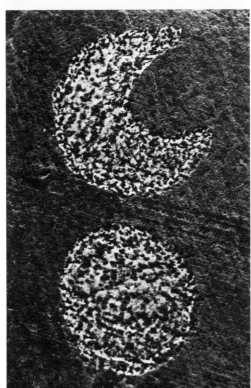

2. Two pictographs discovered in northern Arizona in 1953 and 1954 by William C. Miller. Top: on a wall in Navaho Canyon. Bottom: cave wall in White Mesa. They may be symbolic representations of the crescent moon and A.D. 1054 supernova (Courtesy of Wm C. Miller)

Indians was in use from the tenth to the twelfth centuries. Two more paintings that are possibly the Crab Nebula have been traced in other parts of New Mexico.

As with the oriental accounts, any attempt to reconcile the crescent-circle rock paintings with the known features of the Crab Nebula runs into difficulties. Only one painting actually shows the crescent in correct relation to the star. The best picture has the horns of the crescent directed away from the supernova, whereas in fact they were pointing towards it, and in another sketch the moon is inverted. Ethnologists have challenged the interpretations as well: the Pueblo Indians did not normally record exciting events, not even the burial of their own houses by volcanic ash in A.D. 1067. It is also known that standardized symbols were used as clan signatures to mark visits to specific places. Finally note that the crescent–star pattern is used in other symbolisms throughout the world: about a dozen national flags include it as a design element with a star between the horns of a crescent. Yet no flag designers seem to have been influenced by the event of A.D. 1054, so why treat the American Indians as a special case?

1.6 The birth of the Crab Nebula

Very searching analyses of the oriental sources have more or less removed the lingering doubts that the Crab Nebula and the guest star of A.D. 1054 are intimately related. As we have seen, all Europe and the Arab lands were suffering a pandemic of writer's cramp, so no mention of it has survived. The possibility that prehistoric astronomers in the Americas sketched the event is certainly intriguing and of considerable interest to archaeologists and ethnologists. However, the case is far from convincing and the pictures in any event contain little of astronomical value.

From the Chinese and Japanese chronicles we can fix the birth of the Crab Nebula at around 2 a.m. (local time) on A.D. 1054 July 4 in the Julian calendar. (We have assumed that the skies on previous evenings were sufficiently clear that the explosion could have been sighted had it taken place earlier.) It reached a maximum brightness of -5 mag, sank to -3.5 mag within twenty-three days. Within twenty-one months it dwindled to invisibility, which was probably magnitude 5 as its location would then have placed it in a sky brightened by the zodiacal light. After A.D. 1056 April 17 it

3. Pictograph in Chaco Canyon bearing striking similarity to those found by Miller (Courtesy NASA/Goddard Space Flight Center and John C. Brandt)

disappeared from view, to await the invention of the telescope in Europe and the start of systematic astronomy.

2 · The telescope takes over

2.1 The discovery of the Crab Nebula

Stars and planets are not the only objects in the sky. There are also the hazy patches of light that astronomers term nebulae. When the telescope was introduced to astronomy in the early seventeenth century no startling increase in the number of known nebulae occurred. Since ancient times sky gazers had observed a handful of cloudy patches in the sky, but almost all of these indistinct fuzzes can be resolved into separate stars by a moderate telescope. In 1610 a French astronomer trained a telescope on Orion's sword and observed the blue-green haze that surrounds its central stars. Simon Marius in 1612 re-discovered the Andromeda nebula, already recorded by the Persian astronomer Al-Sûfi in the tenth century. Within its first hundred years, the use of the telescope just about doubled the number of nebulous patches and hazy sources of celestial light. Most practitioners concentrated on studies of the solar system and stars.

The English amateur astronomer and physician John Bevis had his own observatory at his house, and he set out to produce a new atlas of the stars. In 1731 he discovered what we now call the Crab Nebula, but he did not publicize the event. By 1750 the beautiful atlas, with finely engraved copper plates of the constellation figures, was ready, at which stage the printers went bankrupt. The object in Taurus is marked in this atlas, and although Bevis had the proofs, no edition came out until 1786, long after he had died. On 10 June 1771 Bevis wrote to the Frenchman Charles Messier in order to draw his attention to the Taurus nebula, although it is possible that Messier had actually learned about it somewhat earlier in 1763. His colleague Lalande visited England in that year and met Bevis who showed him the atlas and it is therefore conceivable that Lalande reported the discovery to Messier. Messier acknowledged the discovery by Bevis

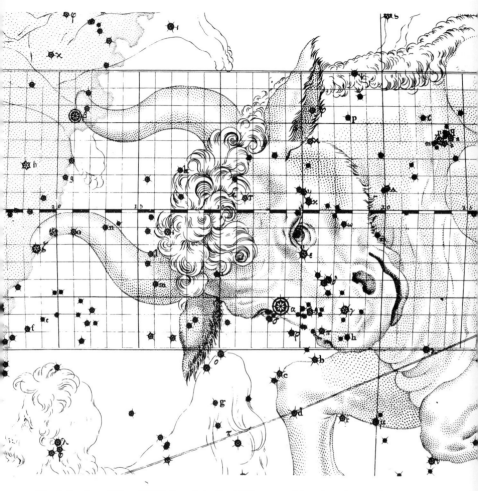

Fig. 3 Portion of Plate XXIII from John Bevis' Uranographia Britannica showing the constellation Taurus. The Crab Nebula, found by Bevis in 1731, is marked as a faint round symbol above and to the right of Zeta Tauri.

in the definitive version of the Messier catalogue published in 1784.

A great astronomical problem of the eighteenth century, quite unrelated to modern astrophysics, resulted in the eventual discovery of the Crab Nebula. Edmond Halley took considerable interest in the bright comet of 1682 and subsequently analysed many observations of comets. As a result of this search he demonstrated that three spectacular comets seen in the years 1531, 1607, and 1682, were all due to a single object moving in an elliptical orbit round the Sun with a period of seventy-six years. On this basis he predicted that the comet would again blaze in the sky towards the end of 1758. He made a plea for astronomers to search systematically for it since he could not hope to live to the age of 102 years and see his prophecy fulfilled. Halley's work on cometary behaviour opened up new branches of research and showed that the laws of gravitation and motion really did give a true account of the orbits. For this reason, and also for the fame bestowed on a discoverer, comet hunting became a popular aspect of observational astronomy.

The outstanding discoverer of comets in the eighteenth century was Charles Messier, credited with sixteen successes. He devoted himself totally to this endeavour, having no wish to grapple with theory and mathematics. In 1758 he commenced the search for Halley's comet which had by then become an observational task of paramount importance: a crucial test of the gravitational theory and the nature of comets. Calculations predicted that Halley's comet would glide through Taurus on successive nights, passing through the horns of the Bull.

Messier's observational report, published in 1759, says this about the discovery of a nebulous object in Taurus on the night of 28 August 1758: '. . . I found the comet of 1758, which ought to be between the horns of Taurus, below the southern horn, a small distance away from the star 'zeta' of this constellation. It appeared as a whitish elongated light spot, resembling a candle in its shape, and containing no stars.' However, Messier was wrong in thinking this nebulous patch of light was a comet. The blur of light did not have the required motion relative to the stars that a comet has. In the end Messier spent eighteen months in a fruitless sweep of the skies. Although he eschewed mathematics he relied too heavily on theoretical work in pursuing the comet. Eventually a peasant discovered it with the naked eye at Christmas 1758, although this news did not reach Paris immediately. Messier finally got his first view in 1759, but by this time the return of Halley's comet was common knowledge, so his claim to independent discovery was treated with derision.

Messier was sidetracked in his searches for true comets by the many stationary nebulae he could see through his telescopes, including that near Zeta Tauri. He decided to draw up lists of these

distractions, and they were published at intervals, commencing in 1771. The nebula in Taurus is the first listed in Messier's catalogue, and is therefore also known as M1. As with Halley's comet of 1758, Messier claimed discovery of the nebula in Taurus, but this is questionable as we shall see.

2.2 Naming the nebula

The publication of Messier's lists stimulated further research on the nature of nebulae. Naturally M1 was included in many of these studies, but they did not contribute significantly to an understanding of it. William Herschel, for example, one of the greatest observers, hoped to resolve the nebula into stars, but failed to do so. His son John Herschel wrote thus: 'As all the observations of the large telescopes agree to call this object resolvable, it is probably a cluster of stars at no very great distance.' The word 'resolvable' here refers to Herschel's belief that it could be resolved into stars with a sufficiently large telescope, and is not an indication that he did resolve it.

The third Earl of Rosse built very large telescopes for observing nebulae and star fields. His research led to the present name of the Crab Nebula. In a paper on nebulae in the Philosophical Transactions of the Royal Society for 1844 Rosse published the drawing that is believed to have led to the use of the term Crab Nebula owing to its superficial resemblance to a crab. Lord Rosse first used this name himself in late 1848. He included the following description of it: '. . . we see resolvable filaments singularly disposed, springing principally from its southern extremity, and not, as is usual in clusters, irregularly in all directions. Probably a greater power would bring out other filaments, and it would then assume the ordinary form of a cluster . . .' Rosse here echoes the then generally held view that the nebulae would clarify into masses of stars with sufficiently high resolving power. Over the years Rosse continued intermittently to observe the Crab Nebula as part of a vast project of nebular observations. In 1880 the Royal Dublin Society published his observations of nebulae and clusters of stars. The remarks on the Crab Nebula underline the difficulties faced when observing nebulae prior to the introduction of photography. For example, the entry for 29 November 1848 reads, 'Crab Nebula. Would have figured it different from drawing in Philosophical Transactions 1844.' Three years later he remarked that his drawing was quite unlike Herschel's.

In 1853 he wrote: 'Appeared very different from the drawing in Philosophical Transactions.'

By 1885 Lord Rosse had enough data for the engraving of a plate of the Crab Nebula, and it is interesting to compare this illustration with modern photographs. Drawings of course have to represent very weak and very intense features within a narrow range of contrast, but nevertheless Rosse's rendition has much in common with the finest photographs. The prominent 'bay' in the eastern part of the nebula is plainly visible in both the drawing and the photographs.

Fig. 4 Sketches of the Crab Nebula made by Lord Rosse. It was Lord Rosse who first used the expression 'Crab Nebula' to describe this object.

One of Lord Rosse's acquaintances was the English amateur astronomer William Lassell, who sought Rosse's advice on the construction of machinery to grind telescope mirrors. With a 24-inch reflector he discovered the largest satellite of Neptune, named Triton, in 1846. Malta's clear skies encouraged him to move his telescope from British drizzle. In his notes on the Mediterranean observations of the Crab Nebula (1852) we find reference to the stars apparently embedded in it as well as to the delicate filamentary structure: 'With $160 \times$ (power of eyepiece) it is a very bright nebula, with two or three stars in it, but with $565 \times$ it becomes a much more remarkable object . . . Long filaments run out from all sides and there seems to be a number of very faint and minute stars scattered over it.'

The general picture of the nebula that emerged from these early visual observations was of an object with complex structure, such as bays and filaments, and with some stars in the field of view. It could not itself be resolved into myriads of stars, as had been the case with the many nebulae now properly called star clusters. The next jump forward in the study of the Crab Nebula was to come with the application of photography and spectroscopy to astronomical observations.

2.3 Into the twentieth century

In 1892 a 20-inch reflecting telescope obtained the first photograph of the Crab Nebula. Later many of the astronomers who were active at the turn of the century took pictures of it. From these early efforts astronomers soon became aware of the unique structure of the nebula, which most thought was an anomalous planetary nebula. Photographs taken in the usual spectral range (360–500 nm*) show an amorphous nebula with a distorted S-shape together with a tangle of bright thread-like features. The nebula has two indented bays on the east and west sides, which give it the S-shape, and there are many variations immediately apparent in the distribution of surface brightness.

V. M. Slipher, of the Lowell Observatory, Arizona, was among the first to secure photographic spectra of the Messier objects. In this connection he is generally remembered for his pioneering observations on the galaxies beyond the Milky Way, research that paved the way to the theory of the expanding universe.

Spectroscopy is a valuable astronomical technique. The light from an object is broken down into its constituent colours. Particular elements are identifiable because they contribute a characteristic set of narrow lines to the spectrum. Stars generally have many dark absorption lines whereas gaseous nebulae display bright emission lines. The spectrum of a sodium or mercury vapour street light contains several strong emission lines.

In the period 1913 to 1915 Slipher took spectra of the Crab Nebula, and in so doing made a significant discovery. He found that individual emission lines in the spectrum are split into two components with slightly different wavelengths. Here is his description and explanation of the effect: 'The prominent nebular emission lines prove to be split into two components which forces us to assume the presence of a Stark effect caused by an electric field.'

*The symbol nm represents nanometres. One nanometre is one thousand-millionth of a metre. This unit is preferred now to the angstrom. There are ten angstroms in one nanometre.

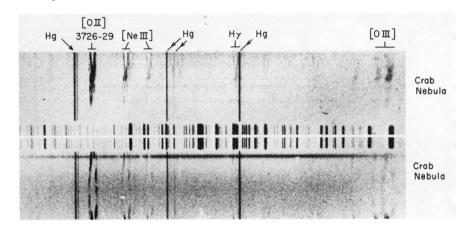

4. Main emission lines in the Crab Nebula have a bow-shape. This splitting of the lines was incorrectly attributed to the presence of high electric fields by V. M. Slipher. For a correct explanation of the effect see the diagram on p.55. (Courtesy Lick Observatory)

Slipher was a great observer and theorist at a time when astrophysics as such had scarcely commenced. He assigned the duplicity of the spectral lines to the existence of a very strong electric field in the nebula.

In 1915 the physicist J. Stark demonstrated in the laboratory that the principal lines in the spectrum of hydrogen atoms split up into several components when disturbed by a strong electric field. Stark used electric fields of 100,000 volts per centimetre, a difficult task at the time. Slipher's conclusion that the Stark splitting occurred in the light from the Crab Nebula is a most interesting one historically for the following reasons. Firstly, the Dutch physicist P. Zeeman had discovered in 1896 that a magnetic field will split up spectral lines into several components; Slipher must have been aware of this work on magnetic splitting. Yet he chose to assign the cause to a newer and less-understood phenomenon, Stark splitting in an electric field, presumably in an attempt to be fashionable. Secondly, this was one of the first occasions on which an astronomer proposed that the Crab Nebula exhibits physical conditions that go far beyond laboratory experiments. Although Slipher turned out to be wrong about the Stark effect, his bold approach was correct: the Crab Nebula is a unique physics laboratory in the sky.

After Slipher's pioneering steps, R. F. Sanford announced further spectroscopic studies of the Crab Nebula in 1919. Some appreciation of the difficulties facing observers at the time can be gained from the fact that he had to expose the photographic plate repeatedly over many nights until the exposure time reached forty-eight hours in order to get his results; the same data would be obtained in a few minutes on a modern 4-metre telescope. Sanford described the spectrum as having two components, one of continuous emission, and superimposed on this, another consisting of bright emission lines. He measured the wavelengths of the latter and correctly identified the atoms responsible for six of the lines.

2.4 The expanding nebula

Dramatic detective work clinching the identification of the Crab Nebula with the guest star of A.D. 1054 came in 1921. C. O. Lampland, working at the Lowell Observatory, compared excellent photographs of the nebula taken over a period of eight years and discovered noticeable motion in individual components. He found that both dark and luminous features had moved significantly in eight years. Even more surprising than this, Lampland found changes in the distribution of brightness over the nebula that could not be attributed to the movement of its separate parts. The nebula seemed to be on the move with parts of it switching on and off!

Among the conspicuous changes he noted were dramatic variations near to the central pair of stars, with several patches of nebulosity altering their shape and brightness. These findings so impressed Lampland that he cabled his results to Harvard College Observatory on 7 March 1921, pointing out that the Crab Nebula was behaving in an entirely unique fashion. Astronomers knew of variable nebulae (e.g. NCG 2261, Hubble's variable nebula) but nothing resembling the commotion in the Crab's tangled web had been found before. Important independent confirmation of Lampland's dramatic result came from John C. Duncan of the Mount Wilson Observatory who measured the relative movement of individual knots and clumps in the nebula. He used two plates, taken with the 60-inch Mount Wilson telescope, separated in time by 11.5 years. This research demonstrated that condensations were moving out radially from the centre of the nebula. It immediately became clear, by backtracking from these motions, that the expansion of the Crab Nebula must have begun about 900 years earlier.

By a curious coincidence three highly important research papers on the Crab Nebula all appeared in 1921: a note by Lundmark on the proximity of the nebula to the position of the A.D. 1054 guest star; Lampland's discovery of the variability; and Duncan's research on the expansion. Today we would expect such serendipity to lead to a great surge in research activity as investigators board a rapidly-accelerating bandwagon. However, in the relaxed academic atmosphere of the 1920s these crucial findings scarcely hinted at the enormous explosion in interest in the Crab Nebula which was to follow in the 1960s and 1970s. The principal findings of these golden years for high-energy astrophysics are the main theme of the remaining chapters. But before starting this exciting story see if you can observe the Crab Nebula for yourself.

2.5 How to see the Crab Nebula

Messier discovered the Crab Nebula in 1758 and Bevis had known about it in 1731. Although it cannot be seen with the naked eye, almost any astronomical telescope is bound to be superior to those that Messier had, which is an encouraging start. Even a good pair of binoculars will bring the nebula into view. However, you cannot expect to see the grotesque filaments and detailed architecture displayed in the photographs. These examples of natural astronomical art are taken with large telescopes having excellent optical qualities. Optical filters are generally used to admit light preferentially from the filaments, thus exaggerating their relative importance. Finally remember that a photograph stores up the light from fainter parts of the nebula until enough has been received to register in the emulsion. Your eyes are unable to improve the optics of your instrument, or to filter out unwanted wavelengths, or to store the image. Therefore you must not expect to see too much.

Before going outside, first of all look at the map to see where the faint smudge of light in Taurus is located. Bright Aldebaran will be easy to pick out if the sky is clear, and from this you can find the horns of the Bull. The brighter star in the horns is Beta Tauri and the fainter one is Zeta Tauri; this third-magnitude star is east-north-east of Aldebaran. One-fifth of the way between Zeta and Beta is the Crab Nebula; it is 1° north and 1° west of Zeta. This is the theory of finding the nebula. Now consider the practical aspects of actually being able to see it.

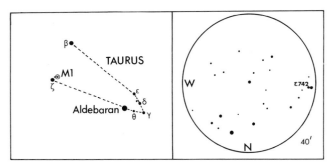

Fig. 5 Finding charts for the Crab Nebula. The sky must be clear and dark with no Moon or else it will be almost impossible to find the nebula. The nebula is 1° north and 1° west of Zeta Tauri and just under ½° west of Σ 742.

The right ascension is 5 hr 31.5 m and the declination 21° 59′ north. Provided you do not live south of latitude − 40°, and hardly anyone does, you have a good chance of seeing the nebula during a few months each year. Let's regard 2200 hours local time as a civilized hour at which to make a search; it turns out that the Crab Nebula is at its highest altitude in the sky for observers north of the equator on 14 January at this time. In practice this means that two or three months either side of the new year will be suitable. The nearer you live to latitude 22° north the less critical the hour of observation. Since the nebula is on the meridian two hours earlier with each passing month the optimum time will be local midnight (2400 hours) in mid-December, 2000 hours in mid-February and so on. A dark night is essential for visual observations of nebulae. It is preferable to observe at least two hours after sunset or two hours before sunrise, unless you are near Earth's equator. There should be no trace of the Moon in the sky and the transparency of the air should be good. This deals with the astronomical constraints, but we must also take human factors into account. The blaze of urban and suburban lighting in many countries and especially the U.S.A. will mean that a sufficiently dark sky will only be found in rural areas. There must be no street lights or illuminated hamburger signs for several hundred metres, although an experienced observer might be able to cope with a less stringent restriction. Finally the observer needs to be dark adapted, with the eye pupils fully dilated; this physical state is accomplished if half an hour has been passed in the darkness of night.

If all the conditions above are satisfied it is just possible to see the Crab Nebula with 50-mm binoculars or any larger aperture instrument with a low-power eyepiece. (Beginners will probably not succeed with 50-mm binoculars except in a really dark sky.) High powers of magnification will make the search harder rather than easier. The detailed map shows forty arc minutes of the field surrounding the nebula, which should aid the search. Look for a faint oval blob of light by sweeping the instrument slowly in the vicinity of the nebula. Generally it is easier to pick up a low-brightness object such as this by sweeping gently because it is then more apparent against the dark background. If you do not have any luck consult an expert for more advice since, like many things in life, it is easier to do after you have been shown once! Of course, these remarks only apply if you are using a relatively simple instrument. For telescopes with setting circles and electric drive it is only necessary to adjust the instrument to the right ascension and declination of the Crab Nebula.

Fig. 6 Tracing of the signal from a simple interferometer telescope showing the interference pattern caused by the radio emission from the Crab Nebula, as it transits the beam of the telescope.

1 hour

More ambitious amateurs may want to try photographing the Crab Nebula, and many people make beautiful colour photographs of it. Experiments will be needed to determine the best exposure time, since this will depend on the telescope aperture, optical efficiency, focal length, and film speed. Accurate guiding of the telescope is essential, however, since the exposure time is certainly of the order of an hour. Finally, you should be aware that unless you have a really outstanding telescope and camera you will not record much more than a smudge of light.

The Crab Nebula is the most intense radio source in the sky, apart from the Sun. This makes it an interesting subject for amateur radio astronomy. Although it is not a variable source, and therefore of less day-to-day interest than the Sun, members of an amateur society can collaborate to build an interferometer that can be used to find the celestial position of the radio source associated with the Crab Nebula. A really ambitious group possessing a receiver having a fast time response might even rise to the challenge of observing the lunar occultations of the Crab Nebula which occur at intervals of several years.

3 · The message of the fiery remnant

3.1 Optical appearance of the Crab Nebula

The systematic investigation of the structure of the Crab Nebula began in the 1930s, when Walter Baade made a series of photographs. He used various filters, each of which admitted only light of a closely defined colour, to bring out the different features of the tangle of gas. Rudolph Minkowski, who collaborated with Baade on many projects, followed this research with an equally careful series of spectroscopic observations. Together they identified the main features of the supernova remnant. They found that it consists of two physically distinct regions: one is an amorphous distribution of gas, responsible for the oval shape, and the other is a chaotic network of filaments. The optical spectrum of the Crab Nebula reflects this structure, for it also has two components: a continuous spectrum, or continuum, superimposed on which is a spectrum of bright emission lines. The continuum, or white light, comes from the amorphous mass of gas that fills out the nebula, whereas the bright lines are generated in the bird's-nest of filaments. The two zones are very different in their physical properties.

3.2 The mysterious continuum

In the historical literature the Crab Nebula is not infrequently categorized as a planetary nebula. Planetary nebulae are not related to planets in any way. They were so called by astronomers such as Herschel because their optical appearance may superficially resemble that of a distant planet in our solar system. These spherical shells of gas in the interstellar medium are each excited at their centre by radiation from a high-temperature star; they are considered to be a late stage in the life of a massive star. Spectroscopy reveals a crucial

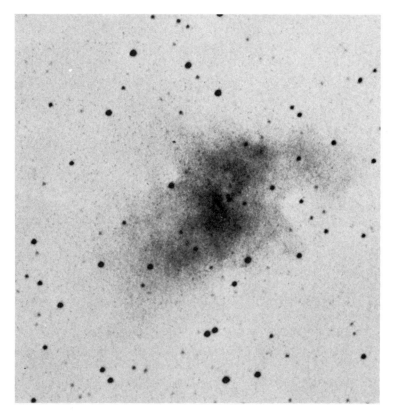

5. The continuum light from the Crab Nebula. (Courtesy Kitt Peak National Observatory, photograph 10208 5)

difference between a typical planetary nebula and the Crab Nebula. For the latter the white-light continuum is extremely intense, accounting for about ninety-nine per cent of the total optical radiation. However, in true planetary nebulae almost the opposite is the case since the continuum is generally inconspicuous, with at least ninety per cent of the energy going into emission lines.

To look at the structure of the source of the white light in the Crab Nebula astronomers have taken photographs through filters in the wavelength range that does not include the strong emission lines. One suitable spectral band is from 720–840 mm, which Baade used to find the morphology of the continuum.

At its outermost boundary the nebula is more or less elliptical, but the intense continuum regions trace out something resembling the letter S in shape. By looking at a photograph of the continuum carefully you can see several features despite the fact that it is frequently described as amorphous. For example, there is less emission from the central region in the neighbourhood of the central pair of 15.9 mag stars, as well as two very obvious 'bays' on the east and west side. Up in the north-west quadrant (n.b. west is on the right-hand side in most astronomical photographs) are regions of varying intensity; it is here that Lampland noticed surprising changes in only a few years. Under excellent observing conditions the continuum source is resolved into a complex mass of thin threads, like wispy cotton wool. The luminous fibres probably reflect the distribution of magnetic field through the nebula.

We do not actually know whether the Crab Nebula is a symmetrical ellipsoid; we cannot tell whether it is prolate (lemon-shaped) or oblate (grapefruit-shaped). Baade found that the major axis measures nearly six arc minutes, and that the ratio of major and minor axes is 0.54, measured to the outermost boundaries. Rotation might give a clue to the three-dimensional structure, but there is no systematic rotation of the nebula about any axis.

The spectrum of the amorphous mass totally defeated the early attempts of theoretical astronomers to explain it. This is because it has such a uniquely simple form. When the intensity of the light is plotted against frequency on a logarithmic scale a steep straight line is obtained with a slope of -2.5. This means that the intensity of light as a function of frequency is governed by a relation of the form: intensity $=$ a constant $\times$ (frequency)$^{-2.5}$. This deceptively simple observation was fiendishly difficult to understand. Objects as diverse as stars, galaxies, gas clouds and planets, which do not have such pure spectra as this, are easier to explain! Let us look into the intriguing astrophysical problem posed by the Crab Nebula.

Suppose we start from the knowledge that the nebula is the ashes of an exploded star. Then we could expect on a simple view that the smouldering ruins should still be fairly hot and glowing for that reason. Then according to detailed physical laws that were derived in the 1930s, the smoothness of the continuous spectrum implies that the gas temperature is high, maybe as much as 100,000 K. But gas does not radiate efficiently at such a temperature. So a vast amount of matter would be needed to do the job of giving the observed brightness. In fact you might need as much as one hundred solar

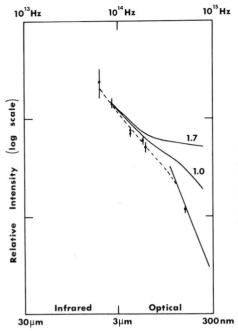

Fig. 7 The continuum optical and infrared spectrum of the Crab Nebula. The curves show the effect of correcting for visual extinction of 1.7 and 1.0 mag, and also without correction.

masses. This seems surprisingly high, since no stars are observed to be that massive so how could the relic of such a star be so big? Furthermore it is hard to see how the web of emission filaments could survive in the harsh environment of such hot gas. Finally, the noticeable structural changes in the Crab Nebula cannot be reconciled easily with the behaviour of other nebulae in the Milky Way that are composed of hot gas.

All these difficulties could have impeded science, but the opposite happened. In fact they inspired the great Soviet astrophysicist I. S. Shklovsky to propose a bold new theory: he envisaged that the strange continuum light did not come from transitions in ordinary atoms in the usual way, but instead from electrons moving through a magnetic field. Two contemporary developments had encouraged him to think along these lines. One was a discovery by American physicists who were experimenting with a synchrotron accelerator. This is a particular type of 'atom-smasher' machine in which electrons are accelerated to nearly the speed of light as they circle endlessly in a magnetic field. At very high speeds the electrons emit blue radiation called 'synchrotron light'. Its production is accounted for by the electromagnetic theory, which shows that when an electron is accelerated—and motion in a circle is a continuous acceleration—it emits radiation. But, unlike an emitting atom, the light comes out as a featureless spectrum rather than a series of characteristic emission lines. This much had been known to physicists since 1912. Then in 1950 the Swedish researchers H. Alfvén and N. Herlofson suggested that cosmic radio sources might be caused by synchrotron emission. The American experiment gave Shklovsky his first vital clue. The second clue came from the sky.

Already in 1953 radio astronomers had measured the radio spectra of several objects and found that plots of the intensity against wavelength on a logarithmic scale gave straight lines. Shklovsky demonstrated that the synchrotron action would power these radio sources, including the radio waves from the Crab Nebula. Then he took the step of proposing the same mechanism for the weird light from the Crab Nebula. This gave the chance of a synthesis of theories: very high energy electrons gyrating madly in the magnetic field produced the strange light of the Crab Nebula, whereas weaker electrons could only manage the radio waves detected from several other objects. In one stroke all the problems of the mysterious light source were solved and they related to the radio emission. Now Shklovsky had to convince the scientific community.

True scientific theories not only explain all the existing knowledge (a mere model or hypothesis does that) but they make predictions of phenomena that will be observed, and thus provide a means for testing their own validity. Shklovsky's notions seemed attractive but dubious: he proposed that the Crab Nebula was a very unusual kind of laboratory containing an atom-smashing machine on an unimaginably vast scale. He desperately needed proof for his revolutionary concept, and to get it he predicted that the light from the Crab Nebula would be found to be partially polarized.

All electromagnetic radiation (such as radio, infrared, optical and X-radiation) consists of fluctuating electric and magnetic fields. Mathematically the radiation can be regarded as a wave motion, with the waves travelling along the line of sight. The radiation is said to be unpolarized if there is no preferred plane of vibration along the line of sight. Radiation is polarized if it is more intense at certain angles along the line. An example may assist an understanding: light reflected from a horizontal surface is polarized in the horizontal

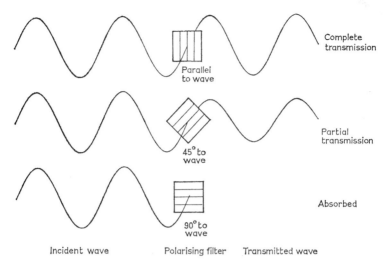

Complete transmission

Parallel to wave

Partial transmission

45° to wave

Absorbed

90° to wave

Incident wave   Polarising filter   Transmitted wave

Fig. 8 Electromagnetic waves are incident from the left onto a polarizing filter. Complete transmission occurs when the plane of polarization of the wave and filter coincide, and complete absorption when these planes are at right angles. Unpolarized radiation has no preferred plane of vibration.

plane. Its intensity (and hence the glare from a shiny surface) is much reduced by a filter (polarizing sunglasses) that blocks light vibrating in the horizontal plane.

Shklovsky's prediction that the Crab's light would be polarized had the merit that it was easy to test: all that was needed was a series of photographs of the nebula through polaroid filters at several angles of polarization. The Soviet observers V. M. Dombrovski and M. A. Vashakidze obtained these independently. At the Byurakan Observatory in the quiet mountains of Armenia Dombrovski measured thirteen per cent linear polarization in the light at the end of 1953. He used a photoelectric detector. Meanwhile, at the Ambastuman Observatory, Vashakidze obtained photographs and immediately discovered that some parts of the Crab Nebula are highly polarized. Baade later took a brilliant series of pictures with the 5-m telescope on Mount Palomar in late-1955. As the polaroid filter is rotated certain parts of the Crab Nebula are completely transformed, some bright parts almost vanishing for certain orientations of the filter! Individual parts of the nebula are over sixty per cent polarized. If the Crab's surface brightness were greater you could see this for yourself by rotating polarized sunglasses in front of a telescope eyepiece.

The discovery of polarization provided concrete proof of the synchrotron radiation mechanism. Here then was a strange new object for astrophysicists to explore: a luminous mass of matter, composed not of atoms, but of electrons and magnetic field. The electrons are travelling at essentially the velocity of light and interact closely with the magnetic field, producing the synchrotron light in the process. Presumably there are protons in the nebula as well but these do not participate significantly in the physical processes that concern us here. Physicists term a material consisting of protons and electrons a plasma. In a sense it is a fourth state of matter, since solids, liquids, and gases all consist of atoms, but in a plasma the atoms have lost their integrity and split up into electrons and nuclei.

The variation of polarization across the object is so extreme and the degree of polarization so high that it is quite out of the question to explain the observations in any other way than by the synchrotron effect. When a moving electron encounters a magnetic environment it is forced to spiral along the magnetic field line. As a direct

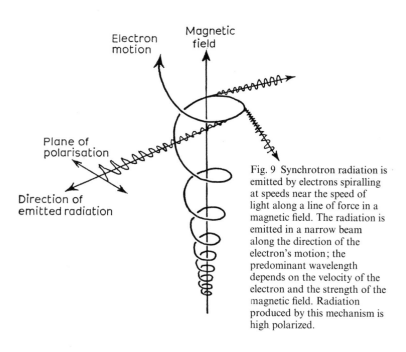

Fig. 9 Synchrotron radiation is emitted by electrons spiralling at speeds near the speed of light along a line of force in a magnetic field. The radiation is emitted in a narrow beam along the direction of the electron's motion; the predominant wavelength depends on the velocity of the electron and the strength of the magnetic field. Radiation produced by this mechanism is high polarized.

consequence the radiation emitted is polarized into a plane at right angles to the magnetic field. If the magnetic structure associated with a celestial object is well-organized high polarization will result; this is the case in the Crab Nebula. The magnetic skeleton can be worked out approximately from the directions of the polarization from point to point across an object. The local field of the Crab seems to run along the edges of the bays, along various wisps, and to wrap around the filaments.

3.3 Continued activity in the Crab Nebula

Shklovsky's immensely important work stimulated renewed interest in the Crab Nebula. For example, the Dutch astronomers J. Oort and Th. Walraven made a systematic exploration of the polarization structure. In doing so they drew attention once more to the astonishing metamorphosis in many parts of the nebula that became evident in a comparison of photographs taken over an interval of several years. In a quantitative comparison L. Woltjer established that individual details had changed in brightness by up to fifty per cent in only fourteen years. Other astronomers confirmed these results beyond any doubt. The most rapid changes in structure occur in the central region, close to the pair of stars, and in the north-west quadrant. Many of the changes cannot be due to actual motion of material. Rather, they give the impression that they are being switched on and off!

Truly astounding information of the activity in the central region came to light in the late-1960s as a result of the research of Jeffrey Scargle. Working on the many plates filed at the Hale Observatories he followed the changes in detail over a period of about thirteen years.

At the centre of the nebula are several variable features which are always strongly polarized and are only plainly visible on plates that register the pure continuum. We can safely conclude, therefore, that these structures are fuelled by synchrotron emission. Certain features of the central area recur on plates taken over a period of years. These are detailed in a diagram showing the configuration of nebulosity relative to the central pair of stars. The most striking feature photographically is here named wisp 1; it is visible on a plate taken in 1910 and is one of the changing wisps noted by Lampland. This particular wisp is present on every photograph of reasonable quality.

Scargle found that wisp 1 seems to oscillate around its average position with an amplitude of a couple of arc seconds and a period of two years. It is a permanent resident of the central region but shifts its position, brightness, and length. Under excellent conditions the wisp itself displays variable structure. According to polarization studies a magnetic field runs parallel to the long axis of the wisp.

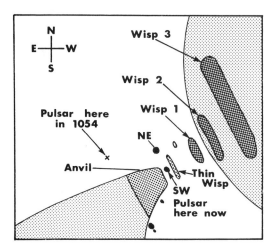

Fig. 10 Major wisp structures and their location in the central part of the Crab Nebula. These wisps are discernible on high-quality photographic prints made from original glass negatives but they do not show up at all readily on reproductions in books and magazines. The position of the pulsar today and its location in A.D. 1054 are also shown.

But this is only one of a whole series of cigar-shaped clouds; others are marked and named in the diagram. Quite often three or more wisps are seen in good photographs. Further from the central stars the wisps get longer and wider. Very close to the south-preceding star, marked S1, is a long narrow streak known as the Thin Wisp, which has never been resolved by telescope. It is strongly polarized and highly variable. Sometimes it vanishes altogether.

In the south-east quadrant of the centre is a region like a trapezoid which Jeffrey Scargle dubbed The Anvil. This also displays the variable wispy structure. Wisps here come and go, and possibly they are generally moving outwards, away from the stars.

A completely new wisp sprang up close to star S1 in September 1969, reaching a length of two arc seconds in a couple of months. This prodigious growth implies that the feature must have raced along at close to the speed of light.

Subtle changes, generally giving the impression that something is rippling out from the centre, occur all over the nebula. All the signs

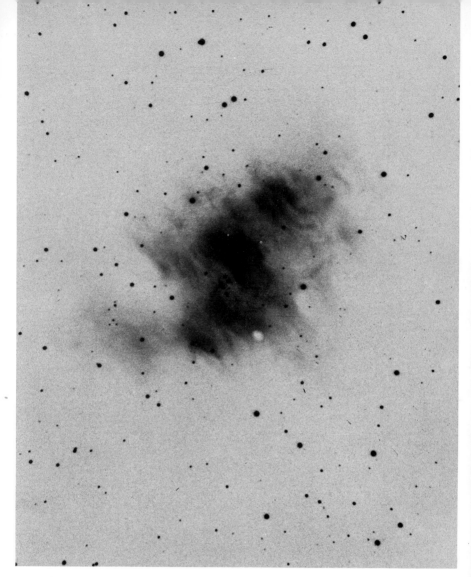

6. A polaroid filter was used in making this photograph of the Crab Nebula in 1967. Note the wispy structures near to the central pair of stars. (Plate PH 5130b, courtesy of the Hale Observatories)

are that the material illuminated by synchrotron light is participating in a general expansion of the nebula. Superimposed on this organized swelling is a more chaotic motion that is most marked at the centre, where movement at up to one-fifth the speed of light is observed.

It would be foolish, of course, to jump to the conclusion that bulk matter itself is hurtling along at these breakneck speeds in the Crab Nebula. Recall that when a stone is tossed into a pond surface waves spread out horizontally from the centre, but the water molecules themselves only move small distances vertically. The high velocities and ceaseless activity in the nebular plasma may not indicate material motion. In the unfamiliar world of the Crab Nebula it is probable that the metamorphoses are made by compressional disturbances, or waves, shooting out from the active zone at the centre. As such a wave moves out it will squeeze together the swarming electrons and the local magnetic field. In a particular small volume of the nebula the electron density therefore rises temporarily and the field intensifies. The electrons wrap themselves more tightly round the bars of their magnetic cage. In consequence the amount of radiation emerging from this compressed region goes up and hence it appears more luminous on photographs. Calculations are very complex, but they show that the emissivity of the plasma rises with the third power of the amount of compression. This behaviour means, for example, that squeezing the electron gas and magnetic field twofold increases the radiation by eight times. In the cores of the wisps the brightness has been intensified by a factor of one hundred, implying a compression factor of about five.

We can sketch a speculative picture of the activity in the following way. Wisp 1 almost certainly oscillates to and fro. To account for this switchback motion rigorously requires a daring plunge into the algebraic jungle of relativistic magnetohydrodynamics—the branch of mathematics that describes the large-scale properties of electrons and magnetic fields. Instead we can look at the problem pictur-esquely. Imagine that somewhere in the centre of the nebula there is a small object emitting very high speed electrons in all directions. For the present we need not consider what this small electron source is. These particles sweep magnetic field and plasma out with them until the energy gets absorbed in the nebular gas at some distance out from the centre. Thus there will be a boundary layer between the hole swept out by electrons and the gas. Wisp 1 is taken as the interface between the hole swept out by the electrons and the surrounding plasma. In a dynamic situation like this it is not possible to maintain exact balance between the outward push of the electrons and the inward pressure of the surrounding magnetic field. In consequence the boundary between the two regimes—wisp 1—vibrates as first one side and then

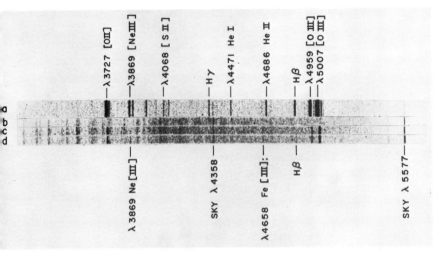

7. Spectrogram of the Crab Nebula with the main lines identified. The
spectra were taken by Roberta Humphreys using the Kitt Peak 2.1 m. The
spectra are as follows (a) Brightest filament in the Crab Nebula. (b) Sky
only near the Crab Nebula, showing many features due to emission in the
Earth's atmosphere. (c) and (d) are spectra of the northern fan jet noted by
S. van den Bergh. (Courtesy Kitt Peak National Observatory and
R. Humphreys)

the other gains the upper hand. This reciprocating motion acts as a
piston that then drives waves and ripples through the fibrous mass of
the rest of the nebula. Note that wisps are seen on both sides of the
nebula, but obviously the real situation is not so simple as in this
model. All the optical observations up to 1969 pointed to star S1 as
the centre of the activity, and later we shall find convincing proof that
this star is indeed the source of relativistic electrons in the Crab
Nebula.

3.4 The fiery filaments

V. M. Slipher obtained the first spectra of the Crab Nebula. These
showed the strong emission lines typical of gaseous nebulae and
planetary nebulae. The emission-line spectrum is certainly reminiscent of planetary nebulae: it is composed of the main lines from
atomic hydrogen (the Balmer series) together with the so-called
'forbidden' lines of ionized oxygen, nitrogen, neon, sulphur, and iron.

Why does the Crab emit forbidden lines? The term forbidden transition is another relic from the early days of physics which arose in the following way. Changes inside atoms are described mathematically by the theory known as quantum mechanics. Simple quantum theory shows that only certain transitions or re-arrangements of electrons are possible in excited atoms. If an atomic electron gets excited to a higher energy state from which it can escape within the terms of the quantum laws it will generally do so in about 10^{-8} seconds, emitting a photon of radiation in the process. Sometimes, however, an excited electron can get jammed into an atomic orbit from which there is no fast escape according to the rules of the quantum game. There are other exits, but these are rather improbable; the excited electron has to sit around for much longer, say 1–1,000 seconds, before being allowed out. The fast escapes are named allowed or permitted transitions, whereas the long-term exits are called forbidden transitions. In the laboratory even the most rarefied gases are sufficiently dense that atomic electrons locked up by forbidden transitions will generally get reorganized in a collision with another atom or electron before they can radiate the 'forbidden' photon. However, the electron density in gaseous nebulae is low enough (less than 10^{14} per cubic metre) for collisional interference to be ruled out, and the forbidden radiation is seen in the spectra of the nebulae.

This now raises the interesting question as to why it is that we see the forbidden lines of oxygen, neon, and so on, rather than the permitted lines of these atoms. The reason for this is that the electron energy levels in the atom that give rise to a forbidden line are not very far above the zero level. Orbital electrons can thus be knocked into these levels by a moderate collision of the atom with a roving electron. With the exception of the hydrogen atom, the permitted transitions are from higher levels that result in radiation in the ultraviolet in many cases, and this is not visible on Earth because it is absorbed in our atmosphere. For example, permitted transitions due to carbon at 154.9 nm and magnesium at 279.8 nm produce strong lines in the ultraviolet, but these cannot be seen with telescopes located on the ground.

The hydrogen lines and the forbidden lines from the Crab Nebula are listed in the table along with their rest wavelengths. In the notation used by spectroscopists the chemical symbol for an atom (e.g. O for oxygen) is followed by a roman numeral whose value less one indicates the amount of ionization (e.g. OII is oxygen that has lost one electron, NeIII is neon that has lost two electrons); finally the square brackets denote a forbidden transition. As an example then, [OIII] 500.7 nm means the forbidden line at a wavelength of 500.7 nm from oxygen atoms that have been stripped of two orbital electrons. Once again we are in a discipline that is laden with historical ballast in the form of a notation that obscures rather than illuminates the underlying physics!

The emission lines are seen to be split up and exhibit a bow-shape in spectrograms of the light from the whole nebula. This is because the nebula is expanding, with parts of it going away from us and other parts approaching. This motion relative to Earth affects the apparent wavelength of the received radiation. Atoms moving away give lines shifted to the red end of the spectrum while those approaching give blue-shifted lines. So the net effect in an expanding object is for light to be shifted to both the red and blue, giving the split lines. In the Crab Nebula the range of velocities, inferred from the spectral lines shifts, is about 3,500 km per second.

The filamentary network of the Crab nebula springs to life in photographs taken through filters that isolate prominent lines and block the white background of synchrotron light. Walter Baade pioneered a systematic approach to this type of work, when he discovered the surprising filamentary detail revealed by a filter that admits the red light of hydrogen at 656.3 nm and the two nitrogen lines that flank it at 654.8 and 658.4 nm. The emission lines come from

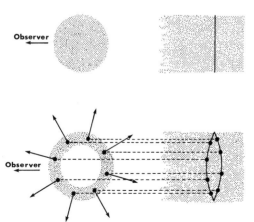

Fig. 11 A cloud of gas which is not expanding has spectral lines with no structure. If the cloud is expanding, however, different parts of it have different velocities with respect to the observer. The effect is to break a single line into a bow structure, as shown. V. M. Slipher incorrectly attributed this structure to broadening in a strong electric field.

a thick shell of fine filaments that encase the amorphous mass of the nebula. Astronomers have made a reconstruction of the three-dimensional structure of this cage of luminous threads. Accórding to Virginia Trimble only a few faint filaments lurk near the centre. Most of them are located two-thirds or more of the way from the centre to the outside of the nebula. Some of the bright features stretch their fingers through the outer half.

Stunning differences are evident in photographs that are taken in various of the emission lines, which implies that there are sharp variations in the spectra of the individual filaments. These spectral variations in turn reflect a variety of physical conditions throughout the tangled network.

The outermost limits of the Crab Nebula in the optical region are defined by the filamentary system, which marks out an elliptic region 7.0 × 4.8 arc minutes in extent. Photographic plates that are exposed until the emulsion is practically burned out show a sharp edge, a result confirmed by more sensitive observations made by image tubes. At the periphery the extended region of emission has a sharp inner boundary also, which is somewhat more intense than the outer boundary. So, although the boundary is well-defined, the maximum intensity occurs about ten per cent of the way in. What is especially significant is that the nebula does not merge imperceptibly with the surrounding interstellar material, but has a definite interface with it.

You can imagine what a tedious job it is to measure up the lengths of this tangle of spaghetti. Donald Osterbrock has done this for the bright filaments and obtained an extent of just over seven arc minutes. This raw measurement must be modified to take account of projection effects, and when this is done it implies a total extent of about 5.3 parsec. We see that these 'threads' are nearly one thousand times wider than Earth's orbit round the Sun! If laid end to end they would go from here to the nearest star and back twice over. On the average the bright filaments are 1.4 arc sec in diameter, which corresponds to a width of 2.5×10^{11} km. The faint filaments would make an even bigger ball of string because their total angular extent is over 700 degrees. If untangled they would stretch for 4.4×10^{17} km, or 20,000 light years—two-thirds of our present distance from the centre of the Galaxy.

Deep exposures of the nebula made at the Kitt Peak National Observatory, Arizona, by Roger Chevalier and T. R. Gull show that thin sheets of emitting gas fan out from the dense filaments. To bring

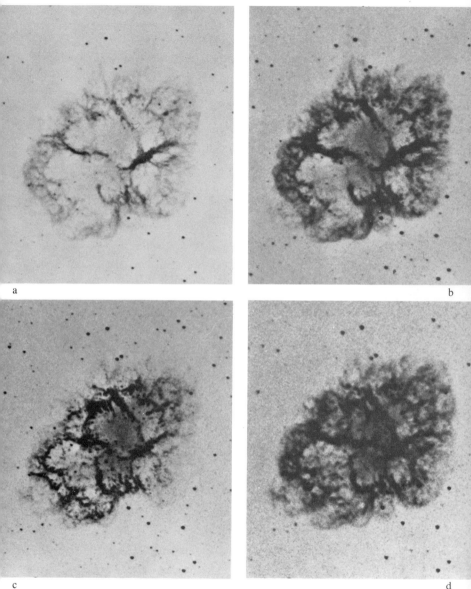

a

b

c

d

8. Four photographs of the Crab Nebula through narrow-band filters reveal a variation in physical conditions of the filaments. (a) and (b) show doubly-ionized oxygen, (c) singly-ionized sulphur and (d) singly-ionized oxygen. (Courtesy Kitt Peak National Observatory, R. Chevalier and T. Gull)

out the sheetlike structures they used filters that transmit only the radiation emitted in one spectral line by material travelling relative to Earth with a small range of velocity, 1,000 km per second. This is less than the total range of speeds in the nebula, which span 3,500 km per second. Therefore narrow-band filters enable astronomers to isolate features moving with a particular velocity and thus to get a cross-section through an expanding object like the Crab Nebula. By using an image intensifier as well as the filters Chevalier and Gull mapped out the sheets of gas, and found that they fan out from the bright ring of filaments on the boundary. One jet of gas on the northern edge of the nebula, first noted by Sidney van den Bergh in 1970, appears to have burst through the tight boundary fence.

3.5 Physics in the filaments

Photographs give different pictures of the nebula in different lines. This teaches us that the physical conditions such as the composition, ionization, electron density and temperature vary from point to point in the nebula. Even a single filament shows 'geological layers' due to variations in these quantities. This means that there is not much hope of close agreement between theory and observation if we only use average values of physical parameters derived from the spectrum of the nebula as a whole.

At first sight it is remarkable that we can say anything about the physical conditions in a cloud of plasma 6,000 light years from Earth on the basis of measuring a handful of spectral lines. Let us look now at the underlying principles of the diagnosis.

The doubly-ionized form of oxygen (two electrons stripped off), OIII, emits forbidden line radiation at 436.3, 495.9 and 500.7 nm, due to transitions in its lowest energy levels. We will call these levels A, B, and C; C represents the highest of the three. When an electron jumps down the atom from level C to level B the attendant energy change results in the emission of light with a wavelength 436.3 nm. Further transitions from B down to A give rise to the lines at 495.9 and 500.7 nm; there are two transitions from level B because the ground state A is actually split slightly into three levels, two of which can be reached by electrons jumping from B. In an object like the Crab Nebula the chances of atomic electrons being knocked into levels B or C and remaining there long enough to make radiative transitions depend mainly on the energy of free electrons in the nebula and only weakly on the density of those electrons. We can see in a qualitative way that the more energetic the free electrons are, the more likely it is that they will jerk an atomic electron from A up to C in a head-on crash with an atom. We can also see that as more electrons are hoisted up to C, as opposed to just as far as B, so the intensity of the 436.3 nm line (C→B) relative to the other lines (B→A) will increase. If the free electrons are not very energetic then maybe only level B will be significantly populated, and the 436.3 nm line will not be seen.

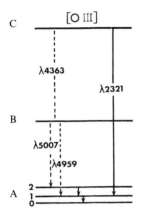

Fig. 12 The energy levels in an atom of doubly-ionized oxygen showing the transitions that produce the lines at 436.3, 495.9 and 500.7 nm. There is also a line in the ultraviolet at 232.1 nm.

In fact the kinetic energy carried by the free electrons is measured by a quantity known as the electron temperature: a general definition in physics of the temperature of a substance is that it is a measure of the kinetic energy possessed by the atoms or electrons that compose the substance. We thus see that the relative population of these levels in the oxygen can act as a thermometer. This is true provided that the density of electrons is not so high that the atoms get out of their excited states (electrons in levels B or C) by means of another atom-electron collision before the time-lag to make a forbidden transition has elapsed. In the gaseous nebulae of the Milky Way, including the Crab Nebula, the density is usually sufficiently low that we can safely ignore reorganization of the atoms in collisions.

For the Crab Nebula the ratio of energies in the 495.9 and 500.7 nm lines relative to the 436.3 nm line is generally in the range 40–80, which brackets a temperature range of 15,000–18,000 degrees. There is an important practical problem in this determination, which arises because the rather weak 436.3 nm line is close to an intense line at

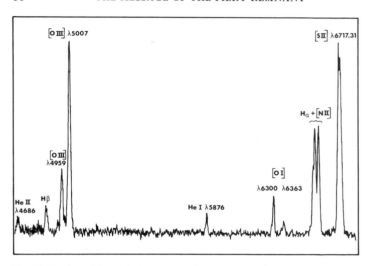

Fig. 13 Portion of the optical spectrum of the Crab Nebula, obtained at the Lick Observatory, after the subtraction of night-sky emission lines. Compare this tracing with the photographic spectrum reproduced as Plate 7.

435.8 nm in the spectrum of mercury. This line is getting stronger and stronger in the night sky as the light from cities continues to grow. Fortunately new spectrum analysers that employ a technique of sky subtraction are now available to remove the contamination at the telescope. Some beautiful spectroscopy has been carried out by Joe Miller, using just such a device at the Lick Observatory.

The problem of determining the electron density is solved by examining intensity ratios from lines that are very close in energy. To take oxygen as our working example again, the singly ionized form has a split energy level above the zero energy state that gives rise to two lines at 372.6 and 372.9 nm. The importance of using two levels with only a small energy separation is that the collisional interference with the two then depends only weakly on the electron temperature, being much more influenced by the density. When the electron density is low (less than 10^7 electrons per cubic metre) most of the collisional excitation is into the lower level, which gives rise to the 372.9 nm line. On the other hand when the density is high (over 10^{10} electrons per cubic metre) the higher level gets more often populated, and this gives the 372.6 nm line. The effect is that a measurement of the precise wavelength of the [OIII] doublet of lines is an indicator of electron density. Osterbrock's measured electron densities in the Crab Nebula vary from 0.5 to 3.7 × 10⁹ electrons per cubic metre. The average value for the bright condensations is 10⁹ per cubic metre.

An idea of the range of physical conditions in the filaments is gained from a temperature calculated from the intensities in the two pairs of sulphur doublet lines at (406.8 and 407.6) and (671.6 and 673.1) nm. This method typically gives an electron temperature of about 7–8,000 degrees, from which we conclude that the sulphur must be ionized under conditions that differ from the ionization of oxygen, where the temperature is twice as high.

A further indication of the variation of temperature throughout the nebula comes from studying the ratio of the energy in the hydrogen alpha line (656.3 nm) to the energies in the neighbouring nitrogen lines ([NII] at 654.8 and 658.3 nm). On average this is large (0.75 or more) for filaments near the centre of the nebula and small (less than 0.25) near the edge. In the average spectrum the value is 0.55. This ratio may show that the temperature increases from the centre to the surface of the nebula. However it could also show that the ionization varies. A similar correlation exists for the neutral helium lines (HeI at 388.9 and 447.1 nm) which also suggest that the electron temperature is highest near the surface. The overall implication is that the Crab Nebula has a source of electron heating that is most effective near the surface of the nebula. Probably this is the result of an interaction between the expanding nebula and the stationary interstellar gas. Besides giving data on the temperature and electron density, the ratios of lines can also lead to a crude derivation of relative proportions of different chemical elements in the nebula. However, these derived abundances of the elements can only be treated as approximations to the real situations, on account of the wide variations in temperature through the nebula and perhaps through individual filaments. When the relative proportions of elements in a gaseous nebula are derived from the emission-line spectrum it is important first of all to have a clear idea of the electron temperature and density. In consequence of the peculiar difficulties posed by the great variations exhibited in the Crab Nebula, rather slow progress followed the preliminary studies by Ludwig Woltjer, published in 1958. He computed the abundances relative to oxygen as a function of assumed electron temperature and the main results are reproduced in the table. Woltjer did an amazingly good piece of detective work that has stood the test of time. He studied plates taken with an old 90-cm telescope, and he did not even have the use of the best plates secured with that instrument.

The range of physical conditions encountered in the Crab Nebula makes it difficult to say how different the composition of the gas is from that of ordinary stars and cosmic nebulae. The early studies seemed to suggest that the Crab Nebula did not differ spectacularly from the normal composition of the interstellar medium for heavier elements. In 1977 Miller stated, on the basis of spectra obtained at the Lick Observatory, that the proportions of oxygen, nitrogen, and neon are more or less normal relative to hydrogen.

Table 3.1

	hydrogen	helium	nitrogen	oxygen	neon	sulphur
T = 8,000 K	37,000	15,000	26	100	76	27
T = 10,000 K	89,000	40,000	254	100	55	31
T = 17,000 K	320,000	20,000	109	100	39	32
Typical Planetary Nebula	170,000	32,000	40	100	15	9

Abundances of the elements in the filaments of the Crab Nebula, relative to oxygen as 100 units, for three values of electron temperature. Values for a planetary nebula are given as well. These data are due to L. Woltjer.

But the real problem is posed by the helium abundance. Helium is the second lightest element and it is manufactured in stars as part of the ordinary energy generating process. Nuclear burning of hydrogen to helium keeps the Sun going, for example. Now what spectroscopists find in the Crab Nebula is that at least half of the gas is helium. Kris Davidson and Roberta Humphreys have found, for example, that up in the northern edge of the nebula the gas mass is 35–40 per cent hydrogen and 65–60 per cent helium. In terms of mass, which is the right way to look at these things, the oxygen is around five times rarer than it is in our Sun. And equally surprisingly other heavy elements seem to be scarce as well. Now if the star that blasted into space to make the Crab Nebula was part of the same stellar population as the Sun, which seems very likely, we would expect that its composition of heavy elements should be similar to or exceed the

Table 3.2

Wavelength	ion	Name	Relative Intensity
342.6	NeV	quadruply ionized neon	27
372.7	OII	strongly ionized oxygen	1260
386.8	NeIII	doubly ionized neon	190
388.9	He I	neutral helium	
410.1	Hδ	hydrogen delta	31
434.0	Hγ	hydrogen gamma	61
436.3	OIII	doubly ionized oxygen	19
447.1	He I	neutral helium	28
468.6	He II	ionized helium	68
486.1	Hβ	hydrogen beta	100
495.9	OIII	doubly ionized oxygen	392
500.7	OIII	doubly ionized oxygen	1192
575.5	NII	ionized nitrogen	11
587.6	He I	neutral helium	79
630.0	OI	neutral oxygen	120
636.4	OI	neutral oxygen	33
654.8	NII	ionized nitrogen	136
656.1	Hα	hydrogen alpha	316
658.4	NII	ionized nitrogen	410
671.6	SII	ionized sulphur ⎱	924
673.1	SII	ionized sulphur ⎰	

Data obtained from photographic spectra by Woltjer show the relative intensity (on a scale with Hβ = 100) of emission lines. Note the great intensity of the forbidden lines from OII and OIII.

Sun. This is a genuine mystery. Its resolution may lie in a new understanding of the origin of the Crab Nebula or in a more complete understanding of the physical conditions within the filamentary web.

3.6 A model of the filaments

We are now in a position to gather up the various fragments of information and present a fairly unified picture of the fiery tentacles. Considerable variations occur within filaments and between filaments. The run of electron temperature is from 8,000–27,000 K, and of electron density from 3×10^8–$3 \times 10^9 \mathrm{m}^{-3}$. The spread in values is due to real variations in conditions both through the nebula and within individual filaments. One task of theory is to explain how the range of electron temperature and ionization has been produced. Apart from an anomalously high helium content the composition is not greatly different from the Sun.

How can we relate information of the filamentary network to the fibrous amorphous mass producing the continuum radiation? The solution to this question lies in looking for the energetic source of excitation in the chaotic and fiery filaments. We already know that the white light continuum radiates through the synchrotron mechanism and that this gives a 'power-law' spectrum (straight line on a logarithmic plot). Suppose we extend this simple law to the ultraviolet part of the spectrum. Then we can infer that the nebula radiates strongly in that part of the spectrum also. This provides the key to understanding the filaments and their emission lines. In almost all cases gaseous nebulae are excited by ultraviolet radiation: in planetary nebulae an extremely hot central star is the power station while in the mysterious quasars even black holes may be involved. The ultraviolet photons transfer energy to the gas and this energy is then radiated in the visible part of the spectrum as emission lines. The process works as follows.

Atoms of hydrogen provide the basic mechanism for extracting energy from the ultraviolet. To take a hydrogen atom to pieces—to separate the central proton from the orbiting electron—requires an energy of 13.6 electron volts.* Photons with a wavelength shorter than 91.2 nm carry an energy greater than 13.6 electron volts. So, along come the ultraviolet photons created in the Crab Nebula synchrotron. They encounter hydrogen atoms. There is an interaction between the atoms and the radiation. Although 13.6 electron volts are used in ripping the atom apart, quite often some energy from an incoming photon is left over since an exact match of energy is seldom possible. This excess energy goes into the motion of the electron (i.e. kinetic energy) in the now-separated atom. Collisions between the photoelectrons, as the electrons knocked out by photons are called, and between electrons and ions, result in a

---

*An electron volt is the energy acquired by an electron when it is accelerated through a potential of one volt. Another way to look at this is that a flying mosquito has a kinetic energy of a million million electron volts.

velocity distribution among the electrons that is typical of the velocity distribution among the molecules or atoms in a gas that is in thermal balance. Thus the electrons, released and energized by the ultraviolet radiation, behave as a thermal gas in certain respects. We can thus define an electron temperature for the gas which is a measure of how agitated, or how energetic, the electron motion is. These electrons, charging about with energies of a few electron volts crash into ions of oxygen, neon, nitrogen and so forth, and knock them into the partially stable energy levels close to the ground state. Downward transitions in these ions radiate the forbidden lines. Thus we can link the transfer of energy due to synchrotron emission from the continuum to the production of the sharp lines from the filaments. In a later chapter this story will be pushed back a few more steps as we discover the source of energy for the synchrotron machine.

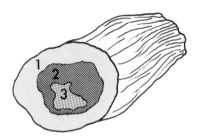

Fig. 14 A schematic cross-section of a filament in the Crab Nebula. (1) High-temperature zone containing ionized hydrogen and helium. (2) Zone with neutral helium, ionized oxygen and ionized nitrogen. (3) Non-equilibrium zone with mainly neutral material. It is hard to find the mass of material in zone (3).

Of course we cannot be absolutely sure that this synchrotron 'engine' scenario is right. But it is a flexible way of getting the range of ionization that is actually observed. Good agreement is found at a surprisingly detailed level between the data and the model. The very strong lines from neutral oxygen, ([OI]) and singly-ionized nitrogen and sulphur ([NII] and [SII]) are never seen in planetary nebulae of ordinary ionized hydrogen regions; these entities are powered by very hot stars. The Crab Nebula must have a different power-station producing the unusual conditions needed to generate these lines. The synchrotron machine gives just the right conditions: photo-ionization from a synchrotron radiation spectrum. Further confidence in the model is assured by the fact that it works so well in many other 'pathological' objects, such as radio galaxies, where extremely strong emission lines are also seen.

3.7 The mass of the Crab Nebula

To help discover the origin of the Crab Nebula we need to determine many things, one of which is to deduce how much material it contains. Are we dealing with the tenuous outer layers of a star, now flung deep into space, or with a mass of material equal to many times that of our Sun? Clearly the answer to this question is crucial to theories of the Crab Nebula.

There are a number of ways to get a mass estimate, and the values do not agree with each other because none of them measures the mass of the whole nebula. A brute force method of getting the mass of the filaments is to multiply the density by the total volume. The electron density is of the order of 10^9 per cubic metre. We can guess that there are three free electrons for every two ions (atoms that have lost electrons) since we see ions that have lost both one and two electrons. To get the volume is easy: the total length of the filaments is multiplied by their cross-section. From these calculations emerge values of 0.02 solar masses for the bright filaments and up to 0.1 solar masses for the faint filaments. However, this sum refers solely to the ionized material and tells us nothing about the vast mass locked in neutral atoms. Nor does it indicate the mass of the invisible gas trapped in very dense filaments.

A better method of getting the filamentary mass is by measuring the energy emitted in the red light of hydrogen gas (H α). This requires knowledge of the electron density, which is given in fact by the ionized oxygen lines. The method gives a range of 0.6 to 3 solar masses for the filaments.

In 1968 Rudolph Minkowski revised the mass of the material producing the emission lines upwards to obtain one solar mass. Subsequently (1970) Kris Davidson and Wallace Tucker argued that the filaments have large amounts of neutral material in their centres, and they speculated that the total mass could possibly be as high as ten solar masses. Miller set the minimum mass of the visible filamentary nebula at 1.5 solar masses. This he felt to be an absolutely lower limit since it made no allowance for non-visible material trapped inside the denser and opaquer parts of the filaments.

Radiation in the forbidden line of neutral oxygen—[OI] at 630 nm—yields another mass estimate. This line is produced in regions where the hydrogen also is neutral. The amount of oxygen necessary to produce the observed intensity across the entire nebula is calculated as a function of assumed temperature. The final step depends crucially on the abundance of oxygen relative to hydrogen. Taking one oxygen atom to every 1,600 hydrogen atoms and one helium atom for every hydrogen atom, Virginia Trimble deduces a range 2.6 to 26 solar masses for a temperature range of 10,000 to 5,000 K. The best guess from this procedure is around 10 solar masses, but it depends on the assumed ratio of oxygen to hydrogen, and assumes equilibrium conditions that have not yet been reached. So it is unreliable in fact.

The space between the filaments, which is filled with the plasma emitting the white-light continuum, has a negligible mass. The Soviet theorist Shklovsky showed that the luminescence from the high-energy electrons in the synchrotron zone, together with the protons to which they were presumably once attached, contributed a mass of 10^{20} kg. This is less than 0.0001 the mass of Earth and only 10^{-9} of the lower mass estimates for the nebula. Now we have a graphic illustration of the importance of knowing the emission mechanism before calculating the mass: the amorphous mass is the seat of almost all the energy emerging from the Crab Nebula at optical frequencies, and yet it contains a negligible fraction of the nebular material.

In conclusion of this study of the classical optical observations we can only state that the total mass of the Crab Nebula and its central star is 1–10 solar masses, with a best guess of 2–3 solar masses. To improve on this broad spread astronomers need spectroscopic data of higher quality than those used so far. It is crucial to make reliable models of the structure of several filaments, and this can only be done with good spectra. Then these improved models can be used as the basis for revised mass determinations. Another vital experiment is the confirmation of the oxygen abundance since this enters as a high multiplier in the computation of the neutral mass.

Table 3.3

Masses of components of the Crab Nebula

Central star (pulsar)	0.5 to 1.5 solar masses
Emission-line gas and filaments	0.6 to 3 solar masses
Neutral gas	Unknown; up to several solar masses possible
Total range	1 to 10 solar masses
Likely range	2 to 3 solar masses

3.8 The nebula expands

Observations published in 1939 by J. C. Duncan demonstrated beyond doubt that the Crab Nebula is radially expanding. By backtracking the velocities of individual filaments we find out when the expansion began. Walter Baade did this in 1942 and came up with the surprising result that the fleeing filaments converged in about 1180, over a century after the sighting of the guest star. He found that the angular speed across the sky of emission knots at the end of the major axis is 0.235 arc seconds per year. This is not as large as the highest proper motions measured for stars (e.g. Barnard's Star is dashing across the sky at 10 arc seconds per year) but it is a dramatic rate of change by nebular standards, especially as the Crab Nebula is so far away. Barnard's Star would only manage 0.01 arc sec per year at that distance.

Virginia Trimble addressed the problem of the expanding supernova envelope for her doctoral thesis in the late 1960s. She measured six photographic plates taken through filters to isolate the emission lines of hydrogen alpha and ionized nitrogen (Hα and [NII]). The photographs came from the large Mount Wilson (2.5-metre) and Mount Palomar (5-metre) reflectors and spanned the period 1939–1966. Some were Baade's plates, superlative works of art, made by accurate telescope guiding under very good skies. Recent photographs do not reach the high standard set by Baade—a man of great skill and experience. Trimble selected 259 filaments and computed the average yearly motion in seconds of arc of each one by comparing the various plates. This detective work showed the date of convergence to be about A.D. 1140, with an uncertainty of ten years. This result left a discrepancy of ninety years or so with the oriental record. Her value for the annual motion of the material at the extremities of the major axis is 0.222 arc seconds per year. The method used is more reliable than Baade's which gave a slightly higher value. The centre of the expansion relative to the south preceding star in the central pair is 10.5 seconds of arc later in right ascension and 5 seconds lower in declination. The 1950.0 celestial co-ordinates of the expansion centre are: right ascension 5h 31m 32.2s and declination 21° 58' 50".

The new measurements of the plates completely confirmed the earlier indication that the Crab Nebula is expanding from a single point. The uncertainty in the convergence date is sufficiently small to

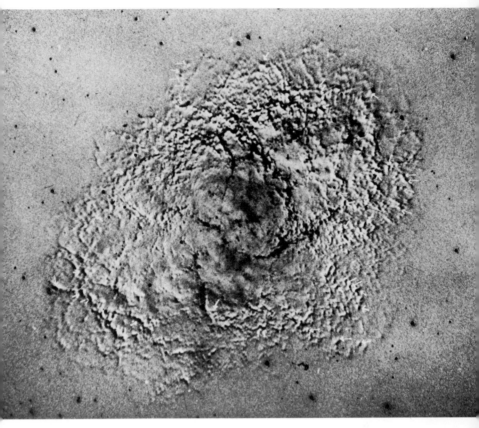

9. The expansion of the Crab Nebula is dramatically revealed in this composite print by Virginia Trimble. A positive print of a plate taken in 1950 is here combined with a negative print from a 1964 plate. Fast-moving filaments have a black leading edge. (Courtesy Hale Observatories and Virginia Trimble)

rule out inaccuracies as the source of discrepancy between the geometrical and historical dates for the birth of the nebula. Instead we must look for an astrophysical explanation of this puzzle. The surprising part of the result is that the convergence date is *later* than A.D 1054. This implies that the filaments are travelling *faster* now than their average speed over the last nine centuries. Something is accelerating the gaseous debris. Had it been the other way round, with the ashes travelling more slowly at later times, the result could be readily understood since a deceleration will occur as the expanding envelope ploughs into the stationary interstellar gas. But in the case of the amazing Crab Nebula some internal 'engine' is clearly exerting a pressure that is more than the minimum required to push back the interstellar medium. The explanation of this expansion energy takes us into the realms of modern astrophysics; it cannot be explained by obvious classical concepts such as the radiation pressure from the light of the central stars.

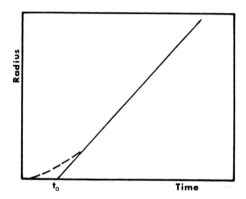

Fig. 15 Schematic illustration of the time behaviour of the Crab Nebula expansion. The presently-observed expansion rate gives an age that is too young. This indicates that the nebular expansion must have accelerated in the past.

Now we are in a position to work out the distance to the Crab Nebula. A distance to an astronomical object can be found if its angular velocity across the sky and true space motion can both be determined. For the delicate filaments we know the angular velocity from the proper motion study and the radial velocity from Doppler shifts in the spectral lines. However, we do not have this information for identical parts of the nebula. Suppose the nebula were a uniformly expanding sphere. Then we should have no trouble because we could just relate the maximum values of the angular velocity (measured at one edge) and the radial velocity (measured at the centre). However, the object is definitely squashed. Therefore the filaments in the outer shell are not all moving at the same rate: those along the minor axis are more torpid than those on the major axis. There is no way of resolving this inherent complication since we cannot yet disentangle the threads of the nebula to reveal its true structure in three dimensions. The consequence is that the largest spectroscopic velocity (1450 km/sec) and the largest angular velocity (0.222 arc seconds per year) bracket the distance at 1.38–2.02 kpc (five to six-and-a-half thousand light years). Other arguments that do not depend on the expansion data point to a value of 2 kpc rather than 1.4 kpc. In this book the value of 2 kpc is adopted since this one receives the most support.

Knowing the distance we can calculate the absolute magnitude that the exploding star attained in A.D. 1054. We have already seen that the maximum apparent brightness was − 5 magnitudes. To find the absolute value we must work out how bright the star would have looked at a distance of 10 pc as opposed to its actual distance of 2 kpc. This procedure is dictated by the definition of absolute magnitude, namely the apparent magnitude to an observer 10 pc away. The ratio of the distances is 200, leading to a difference of 4×10^4 in the ratio of brightnesses. When we transform this information to the physically-irrational system of units still employed by astronomers we get an absolute magnitude − 16.5 mag. One more correction must now be made. Interstellar matter intervenes on the 6,500 light-year path to the nebula. From optical studies of stars near the Crab Nebula it seems that interstellar dust dims the light from stars near the Crab nebula by 1.5 mag. Putting all the information together yields an absolute magnitude of − 18 for the guest star. This is an immense value, comparable in fact to the energy radiated by an entire galaxy. Our Sun has an absolute magnitude of about 4.5. Briefly the star of A.D. 1054 equalled the combined light of 500 million stars like the Sun. Such astonishing profligacy immediately demolishes any remote possibility that the A.D. 1054 pyrotechnic display was a routine nova explosion. Without doubt the orientals recorded a rare astronomical event: a supernova explosion, marking the death of a massive star and the birth of the beautiful remnant visible today.

3.9 Collision with the interstellar gas

The ejecta from the Crab Nebula, rushing outwards at thousands of kilometres every second, have a profound effect on the ambient gas and dust in its immediate vicinity. The electron temperature increases towards the edge of the nebula. The collision of the expanding envelope with the stationary interstellar medium probably provides the extra energy needed to heat up the edge. As the supernova remnant grows it sweeps up the interstellar medium before it, and thus increases the total mass of the nebula. A significant deceleration of the expansion does not take place until the expanding shell has snow-ploughed something like its own mass. The material in the Crab

Nebula has not yet had time to crash its way through its own mass of interstellar material. It's rather like trying to stop a runaway vehicle with a heap of wood shavings—you have to have a very big pile to have any effect. In this process the kinetic energy of the remnant is gradually donated to the interstellar medium. The collision of supernova shells with the ambient gas provides one method of stirring up and heating the interstellar medium in our Galaxy. Other heat sources probably include cosmic rays and stellar winds. Supernovae contribute significantly to the 'weather' between the stars, by stirring up and heating the interstellar 'atmosphere'.

Calculations of the detailed interaction of the Crab Nebula with the interstellar gas are naturally complex. Equally it would be hard to model the exact behaviour of those woodshavings dashed aside by our imaginary vehicle. We can be reasonably certain that the interface between the two is a region of considerable turbulence: a hot dense gas cannot be projected into a cold less-dense gas without producing considerable turbulence. The chaotic conditions existing at the boundary may provide a mechanism for accelerating heavy particles (ions) in the debris to velocities close to the speed of light. In this and in other ways the expanding cloud may contribute to the density of cosmic ray particles in the Galaxy.

3.10 Location in the Galaxy

The Crab Nebula is situated at galactic longitude 184° and latitude −6°. This places it further out from the galactic centre than the Sun, and 200 parsec below the main plane of the Galaxy. Its motion is rather uncertain, due to the difficulty of extracting reliable information of the proper motion of the point from which the expansion seems to have originated. However, the object does not deviate strongly from the circular motion round the Galaxy, as exhibited by our Sun, for example. In the 1960s an idea that gained currency was that the Crab Nebula resulted in the explosion of a runaway star. Very high velocity stars sometimes escape from the families known as star clusters or associations. One imaginative hypothesis envisaged that the precurser of the Crab Nebula had been slung out of the 1Geminorum association in a slingshot interaction with other stars. However, Virginia Trimble's data on the motion of the nebula, combined with other of its properties, appear to rule out this interesting possibility.

3.11 The theoretical task

Finally here is a summary of the known optical properties of the Crab Nebula. This overview gives a clear idea of just what theoretical astronomers have got to account for and to explain away.

The nebula is composed of two distinct parts. In the ghostlike amorphous region, which is composed of a multitude of thin threads, emission is caused by the synchrotron effect as electrons travelling at the speed of light weave a path through the magnetic field. Encasing the synchrotron continuum is a dense system of filaments. Like street lights and gas clouds these have a rich emission-line spectrum, indicating a range of electron temperature 8,000–27,000 K. The mass of the nebula is 1–10 solar masses. Changes in the optical appearance betray high activity in the vicinity of the central stars. It is expanding at roughly 1,500 km/sec, and the outward motion is accelerating. There is nothing remarkable about its location in or motion through our Galaxy. It is 1.3–2 kpc from us, which leads to an estimate of − 18 mag for the absolute magnitude of the initial explosion.

The first task for theoretical astronomy is to work out the energy budget and the energy transfer mechanisms within the nebula. Before embarking on this challenging problem, however, it is essential to look at the properties of the nebula in the invisible parts of the electromagnetic spectrum.

4 · The invisible nebula

4.1 Electromagnetic radiation and the Crab Nebula

A vast amount of information about the Crab Nebula is gleaned from the optical radiation. But we must throw off our optical blinkers and look at the emission across the entire electromagnetic spectrum. Radiation is detectable from a frequency of 10 kHz, representative of long-wavelength radio waves, right through the microwave, infrared, optical, ultraviolet, and X-ray regimes, and on through to gamma rays with a frequency of 10^{20} Hz. This range of thirteen decades between the extremes of the observable frequencies is practically unique. Certainly the Crab Nebula gave positive results across such a broad electromagnetic span before any other celestial object.

Since 1945 the astronomy of the invisible parts of the spectrum has developed at an impressive pace. Each new technique threw back the curtains and opened a new window on the energetic cosmos. During the development states of new techniques the Crab Nebula was frequently one of the very few objects that could be seen with novel instrumentation. It is relatively bright in several of the wavelength windows, being among the brightest in the sky in the radio and X-ray regions. Paradoxically this contrasts with its relative feebleness in the optical region, where it competes for our attention with brilliant planets, stars that do not emit the invisible radiations very strongly.

The structure of the nebula, with its wisps, filaments, and amorphous mass, is evidently complicated. There is a temptation, therefore, to assume that measurements of the total radiation received from the nebula will be practically useless on account of its many physically distinct parts. This is not the case, however, because the structure of the nebula can be mapped in detail at a selection of fixed wavelengths. These maps can be used to distinguish the relative contributions of the constituent parts. A second point is that measurement of the total radiation is comparatively easy. Indeed in certain cases it provides the only information available.

The range of energy and frequency is inconveniently large when it comes to plotting out the spectrum. Ordinary or linear graph paper, in which lengths along an axis represent equal amounts of energy, wavelength, or frequency, cannot be used because the size of paper needed to display all the details would be absurdly large. For instance, if the spectrum from 10 MHz to 10 GHz were plotted, linearly across this page the rest of the spectrum would need a sheet of paper from here to Saturn! Instead the graphs are plotted on logarithmic scales in which length indicates equal ratios of the quantities plotted. Here we follow the usual practice in radio astronomy and use a scale in which the energy or frequency alters by a power of ten at every unit along the axis. The energy range (vertical axis) covered is about nine decades and the frequency range (horizontal axis) thirteen decades.

Radio astronomers have made many accurate determinations of the energy received from the Crab Nebula, as well as a few inaccurate ones along the way. Putting all the information together in a

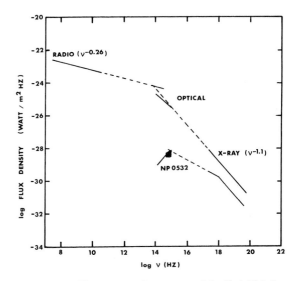

Fig. 16 Electromagnetic spectrum of the Crab Nebula at all wavelengths. Solid lines indicate the observed behaviour. The spectrum of the central pulsar is also indicated.

logarithmic plot shows that a straight line can be chosen to fit the points tolerably well. The slope of the line is -0.26, so the radiation is represented by a relation of the form $S(v)$ is proportional to $v^{-0.26}$, where $S(v)$ is the energy at a frequency denoted by v. The index (-0.26) in this formula is called the spectral index. Incidentally, the energy is measured in units of watts per square metre per hertz (W m^{-2} Hz^{-1}): the watt is the standard unit of energy in the metric system, being defined as one joule per second; the method by which the energy is collected is standardized to a square metre and to a bandwidth of one hertz. These units, while eminently sensible for laboratory physics, are inconveniently large for astronomers who have to measure tiny fluxes of energy. At a frequency of 10 MHz the Crab Nebula has a flux of about 4×10^{-23} W m^{-2} Hz^{-1}, which falls steadily to about 2.5×10^{-24} W m^{-2} Hz^{-1} at 100 GHz.

As noted on p. 44, in the optical range the spectrum is also a straight line in a logarithmic plot, but the slope is much steeper, -2.5, before correction for the effects of obscuration by interstellar matter. The earliest reliable infrared measurements date from 1968 when American scientists obtained data in the near infrared. By 1971 the measurements had been pushed to a wavelength of 10 microns (frequency 3×10^{13} Hz). These points in the infrared fall onto a smooth extension of the optical spectrum. Now we must face a tricky point: how much dimming by intervening dust should be put into the spectrum? The most popular estimates (not necessarily correct!) favour a reddening of one magnitude. If the points are recalculated with this factor everything from wavelengths of 10 microns to 300 nm fits a line with a slope of -0.9.

Up at the high-energy end of the spectrum many measurements, made at truly vast public expense, exist. There are problems in relating all the X-ray and gamma ray observations because of the difficulties encountered in calibrating the instruments used. Nevertheless a smooth spectrum with a slope of -2 describes most of the data satisfactorily.

When all the spectral data are put together it is evident that three straight lines with different slopes give an account of all the information on the total radiated flux from the Crab Nebula. Note how the two points at which the slope must change fall in the unexplored regions of the spectrum. Filling in these gaps is scientifically more important than getting higher accuracy at fabulous public expense in the parts already studied.

From the astrophysical viewpoint the important information is that the spectrum follows a 'power-law' behaviour over such a vast frequency range. Synchrotron radiation is responsible for the characteristic straight lines, and it evidently extends right into the X-ray regime in the Crab Nebula. In the Crab Nebula synchrotron X-rays can only be produced by electrons moving very close to the speed of light, so close in fact that their effective mass is millions of times higher than their rest mass. Where do these immensely energetic electrons come from and how do they acquire their relativistic velocities? Discussion of these vitally interesting questions will continue when we have a clearer picture of the nebula at all wavelengths.

4.2 Radio waves from the Crab Nebula

A discussion of the Crab Nebula's radio emission takes us back to the pioneering days of radio astronomy. In 1948 an Australian research group led by John Bolton discovered four separate radio sources; one in the constellation Taurus they called Taurus A. (This group of astronomers incidentally has never really had fair credit for its early achievements.) The following year the group made a positional measurement, using a rudimentary radio interferometer to give an adequate angular resolution, and tied the position of the radio source down to within seven minutes of arc. Within these limits the radio source position agreed with the Crab Nebula. So Taurus A became the first radio source beyond our solar system to be positively identified with an optical object. In the following years the Australians stayed well in the lead. Bernard Mills of the University of Sydney measured the angular diameter at 100 MHz in 1952 and produced the first crude map of the distribution of radio emission in 1953. This map showed that the gross features of the radio emission had the same shape as the optical nebula.

Cambridge radio astronomers have produced beautiful maps of the Crab Nebula's radio structure by using the Cambridge One-Mile telescope. This instrument consists of three parabolic dishes, one of which can be moved on an accurately surveyed railtrack. It is possible to build up the angular resolving power of a telescope one mile (1.6 km) in diameter by positioning the movable antenna at different points along the track and by observing an object (e.g. Taurus A) for successive twelve-hour periods. The combination of the variable spacing of the telescope elements and the rotation of the Earth means that the movable dish can eventually be placed in all possible configurations relative to the two fixed dishes up to a separation of one mile. A computer has to be used to transform the data into the information that a conventional (but impossible-to-build) single dish would give. In the case of the Crab Nebula the observations took just over one month of observing time, during which the radio source was tracked for twelve hours every day, the movable antenna being pushed 77 feet (23.5 m) along its railtrack between successive observations. The telescope operated at frequencies of 2.7 and 5 GHz (wavelengths of 11 and 6 cm respectively), to give a maximum resolution of six seconds of arc. This is considerably larger than the smallest detail visible in optical photographs taken under good conditions, but the best that can be achieved at present. In keeping with the usual practice in radio astronomy the observations are reproduced here as contour maps, which are exactly analogous to the representation of hills and mountains on a geographical map: the highest contours mark the places of maximum radio intensity.

The map of emission at a frequency of 5 GHz, produced by Andrew Wilson, contains great detail, which is astonishingly similar to the distribution of optical continuum emission—the synchrotron light from the amorphous mass. Among the common features are the S-shaped ridge, which is even more evident in the radio map, the bays on the edge of the nebula, and a small valley near to the north-west. An important difference is that the radio picture of the Crab Nebula is at least fifty per cent larger than the optical extent across most of the brighter parts. However, the outermost limits of the radio and optical pictures coincide remarkably well. A close comparison of the radio map with the brightest filaments shows that some of the ridges in the map definitely coincide with the dense threads of light. Thus, some of the radio emission appears to be associated with the filaments. This may be because the magnetic field is about fifty per cent higher near to a filament, thus enhancing the synchrotron process that is responsible for the radio emission.

The optical activity in the centre of the nebula raises an obvious question: what can a radio telescope see in this interesting region? Within about one minute of arc of the central stars there are three or four ridges of radio emission. At the centre there is a reduction in the radio intensity which corresponds in shape to lower brightness of the optical continuum. Insufficient time has passed since these obser-

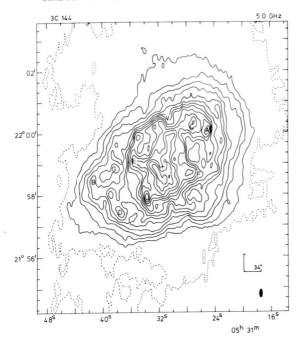

Fig. 17 Radio contour map showing the structure of the Crab Nebula at a frequency of 5 GHz made by the radio telescope at the University of Cambridge.

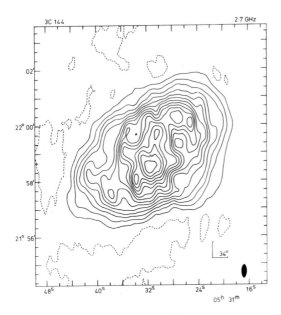

Fig. 18 Radio structure at 2.7 GHz.

vations were made (1970) to say whether the structure at the centre is changing significantly. Wilson tested for this effect on a time base of nearly two years, but found that although the possibility of changes could not be excluded, they were certainly not detectable with the Cambridge interferometer. This is an experiment which should be repeated with the longer time base now available.

Radio maps made at lower frequencies with the same telescope baseline (1.6 km) have poorer resolution. The main interest in examining these smoother maps is to see if the structure changes with frequency. In fact there is scarcely any discernible change in the radio images from a frequency of 408 MHz right up to 86 GHz, apart from a very slight decrease in size. This shows that on scales of an arc minute or so, corresponding to lengths of 2×10^{13} km (one quarter of a light year), the general features of the magnetic field and relativistic electron density do not vary much with position. It is important to remember that the high-speed electrons, responsible for the optical and radio synchrotron radiation, are billions of times more energetic than those that excite the forbidden line radiation and are thus a quite distinct population. According to the radio maps these relativistic electrons are more uniformly distributed than the low-energy thermal electrons which are localized into the filaments.

The synchrotron radiation theory received impressive confirmation when a brilliantly executed experiment by Soviet astronomers found a polarized component in the radio emission. Such behaviour is highly characteristic of synchrotron radiation. Its presence leads to the exciting possibility of discerning the magnetic field in the nebula. Radio telescopes can make maps of the polarized component of the radiation in the following way. At the focus of each parabolic dish there is a dipole (antenna) which responds to the radio signals from space. This dipole may be rotated, and when this is done the response generated by unpolarized radiation is unchanged, but that from polarized radiation depends on the angle between the dipole and the electromagnetic wave. So, if the dipole is at right angles to the plane of polarization no signal is received, whereas if it is parallel a maximum response results. The trick is to make a set of three maps with the dipoles at 0°, 45°, and 90°. By subtracting maps in a suitable way the polarized component is brought out. Similarly, by rotating polarized sunglasses you can work out the polarization plane of light reflected from a smooth surface. When the glasses transmit the least light they are at right angles to the plane.

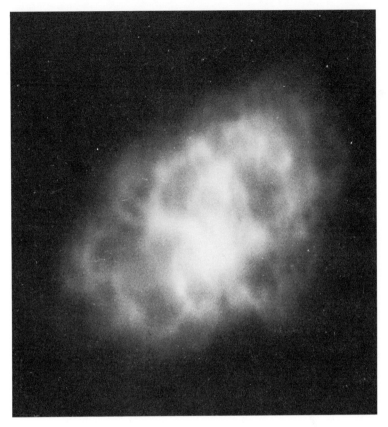

10. The Crab Nebula's radio emission at a frequency of 5 GHz is displayed as a positive 'photograph'. The nebula has a softer outline but the filaments show clearly. (Courtesy University of Cambridge, Mullard Radio Astronomy Observatory)

In the radio domain an interesting effect known as Faraday rotation is encountered. As a polarized wave travels through a part of space containing free electrons and magnetic field the plane of polarization gradually twists round. The amount of rotation depends on the density of electrons and the magnetic field intensity. Since the Crab Nebula is filled with electrons and magnetic field Faraday rotation will have affected the radiation. Furthermore the density of electrons varies from place to place, so the precise amount of rotation changes with position. This Faraday effect plays a secondary role of reducing the polarization of the radiation, because the superposition of waves with many arbitrary rotations arising at different points along a line of sight muddles up—and hence reduces—the overall polarization. The end result is that polarization maps tell us only about average conditions along a line of sight. As is frequently the case in astronomy the information in the crucial third dimension, that of depth, cannot be reconstructed uniquely.

The amount of polarization at a frequency of 5 GHz is lowest in the vicinity of the brightest optical filaments. This depletion is certainly due to variations in the Faraday rotations at different points in the filaments. In fact the vibration plane of the radio waves must be twisted through 360° or so to account for the reduction in polarization. This observation immediately yields a value for the

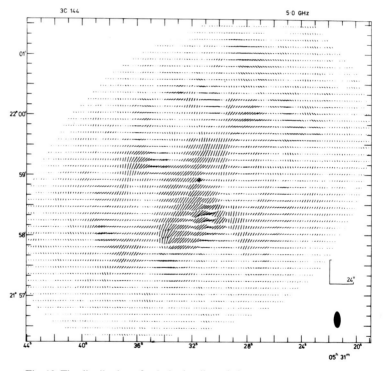

Fig. 19 The distribution of polarized radio emission at a frequency of 5 GHz. The bars are aligned along the preferred axis of the radiation and their length indicates intensity. The average magnetic field runs at right angles to these bars.

average magnetic field inside the tangled filaments. For an average density of 10^9 electrons per cubic metre the magnetic field is roughly 10^{-8} tesla. (One tesla is 10^4 gauss; the magnetic field at the surface of Earth is about 10^{-5} tesla.) Odd threads of gas may have slightly higher magnetic fields. Although the field strength may look rather feeble in comparison to the natural field of the Earth, remember it is thought that the Earth has an internal dynamo to maintain its field, and that the volume occupied by the filaments is on a much vaster scale than our entire solar system. In our Galaxy a typical interstellar magnetic field is 10^{-10} tesla (one microgauss). In the Crab Nebula the field is 100 times higher than this.

From a comparison of polarization maps made by optical and radio telescopes it is apparent that large-scale rotation of the plane of polarization takes place over the whole of the face of the nebula. These gyrations are also due to the Faraday effect, probably caused by the electrons in the innumerable faint threads of gas that can be seen under the best observing conditions.

The story of the radio observations of the Crab Nebula stretches over thirty years. From the melange of data on the spectrum, polarization, and distribution of the radio emission a reasonable picture of the synchrotron power-station that dwells within the Crab Nebula has been built up. Its magnetic structure is defined and it corresponds with optical features such as the bright filaments. Some insight has been gained into the distribution of the many things—thermal electrons, high-energy electrons, magnetic fields, and radiation—that make up the complex world of a supernova remnant.

Radio astronomers had a head start over the other practitioners of invisible astronomy. Consequently less is known about the behaviour of the Crab Nebula in the infrared, X-ray, and gamma-ray regions of the spectrum. This is more a reflection of the vast body of radio and optical data than of any lack of effort on the part of infrared and space astronomers, who have gathered as much information on the Crab Nebula's appearance in these wavebands as for other bright objects.

4.3 The infrared nebula

The infrared universe is even now a neglected pane in the astronomical window on the cosmos. Sprawling over three decades of frequency, from 3×10^{11} to 3×10^{14} Hz (wavelength 1 mm to 1 micron), the infrared region of the spectrum bridges the optical and microwave radio regimes. Telescopes intended specifically for infrared studies will soon roll back the frontiers, but meanwhile there is little to go on in the case of the Crab Nebula. Although it was an obvious object for the pioneering infrared astronomers to look at, it did not finally give a positive result until 1968. Measurements of the infrared emission unite the optical and radio data without any significant jumps. All is consistent with the synchrotron mechanism, even though the number of observations is frighteningly low. The Crab Nebula is one of the few synchrotron sources that infrared astronomers can study. Most of the infrared photons that they collect come from warm dust shells swathing young stars.

4.4 The Crab Nebula viewed from space

X-rays cannot penetrate Earth's protective atmosphere. Astronomers who work with cosmic X-rays have to use detectors and telescopes placed above the atmosphere. Initially balloons and rockets were used in attempts to get a brief glimpse at cosmic X-ray sources, but now permanent X-ray observatories are orbiting the Earth as artificial satellites. Several supernova remnants are X-ray sources, among them the Crab Nebula.

In the early 1960s incredible scepticism surrounded the suggestions that any money whatever should be spent on looking for a few cosmic X-rays from beyond the solar system. The first X-ray project launched at public expense in the U.S.A. officially had the quite extraordinary brief of looking for X-ray luminescence from the Moon's surface. This project has all the appearance of a smokescreen to get bureaucrats to fund something they would understand (looking at the Moon) in the hope that genuinely useful science (looking at rubbish from dead stars) would be done as well. Whatever the truth of the matter, on 18 June 1962, to satisfy the rules of the funding game, the rocket made a nominal look at our nearest neighbour and saw nothing remarkable. However, during its flight lasting a few minutes it did detect a strong source of X-rays near the Moon as the rocket rolled in space. This extraordinary object, dubbed Scorpius X-1, fully justified all the hopes of X-ray astronomers, for its enigmatic properties soon opened the public purses for more rockets and detectors. In this financially and scientifically buoyant atmosphere the Crab Nebula was a sitting

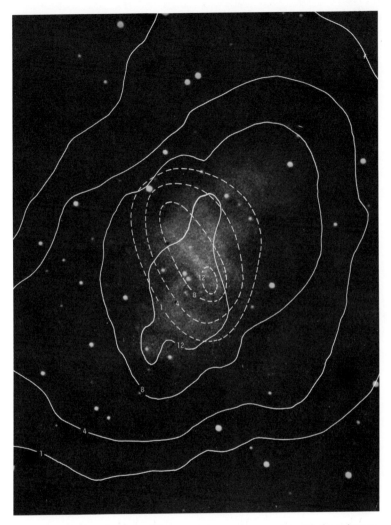

11. A superposition of radio and X-ray emission contours on the optical continuum of the Crab Nebula. The solid lines are 5 GHz contours from the work of Wilson. The dashed curves conceptualize the 1–4 keV X-ray source associated with the pulsar. (Courtesy University of California, Lawrence Livermore Laboratory and US ERDA)

target because theorists had predicted that it would be found to be an X-ray source. Sure enough in 1964, it was discovered to be radiating about 10^{30} watts in X-rays, and thus it became the first X-ray source outside the solar system to be positively identified with an optical object.

Early experiments in X-ray astronomy were crude. The first generation of X-ray telescopes did not have very good ability to determine the structure of the Crab Nebula. However the structure and location of the X-ray source in the Crab Nebula was crudely determined in 1964 during a lunar occultation. Such lunar occultations occur periodically when the Moon crosses between Earth and the Crab Nebula. On these occasions it is possible to determine the X-ray structure by following the change in the received X-ray signal as the edge of the Moon glides across the face of the nebula. A series of Moon–Crab Nebula occultations occurred in 1964 and 1974.

The measurements made in 1964 showed two important facts; that the X-ray source at energies of 2–10 keV is extended, with a size of around two minutes of arc, and that the centre of the emission coincides with the central pair of stars. Subsequently other rocket flights confirmed the general picture of an X-ray source in the Crab Nebula that more or less mirrors the distribution at optical and radio frequencies.

Excellent opportunities to study the spatial distribution of X-rays arose in late-1974 when a new series of lunar occultations commenced. Several groups of scientists got well-prepared to send detectors to a high altitude in balloons and on rockets. These experiments revealed that the X-ray source in the Crab Nebula is less extensive than the optical continuum emission. Although the centroids of the two distributions match up, there is an offset of about 10 arc seconds between the maximum X-ray intensity and the centre of activity in the nebula. Observations over a range of X-ray energies demonstrate that the observed size of the nebula shrinks at higher energies. In the range 0.5–16 keV the total diameter is about 80 arc seconds; this falls steadily with increasing energy, reaching 50 arc seconds at 50 keV. The shape is elongated and the dimensions here refer to the longer axis. The long axis of the X-ray structure runs in the NE–SW direction, which is at right angles to the major axis of the optical nebula. There is some evidence that the bulk of the high-energy X-ray emission is in fact associated with the series of wisps in the active centre of the nebula.

X-ray maps to the standards set by radio astronomers will not be available for some time. They require a telescope with an excellent imaging system, such an expensive system that it will have to be permanently in orbit, rather than ditched in the ocean after one rocket experiment, to justify the cost. Three satellite observatories (HEAOs) for high-energy astronomical work will be launched in the 1980s, to probe the universe of general cosmic violence.

In addition to the structural work X-ray observers have also found that the Crab Nebula's radiation has a polarization of around fifteen per cent in the X-ray domain, confirming yet again the existence of the synchrotron mechanism in the nebula. All the X-ray results can be accounted for in terms of the photon screeches from very energetic electrons as they wrap around the ambient magnetic field.

Measurements of the X-ray spectrum are numerous and they can be accommodated by the normal power-law relation with a spectral index of 2. The intensity does not change with time, at least down to the five per cent level. Searches have also been made for emission lines in the X-ray spectrum due to transitions in highly-ionized atoms but none have been found.

Above a photon energy of 1000 keV (1 MeV) we are in the gamma-ray territory, where the number of photons expected from any celestial object is low simply because each photon delivers a large packet of energy. Once again the Crab Nebula was among the first of the discrete objects to reward investigators and it was the first gamma-ray source identified with an optical object. The gamma-ray spectrum fits smoothly onto the X-ray counts with a spectral index in the vicinity of -1.2. No structural information is available at the higher energies.

4.5 Relativistic electrons and magnetic fields

At every frequency of observation the bulk of the radiation from the Crab Nebula is accounted for by the synchrotron radiation mechanism. At ultraviolet, X-ray and gamma-ray frequencies, which is where most of the energy is radiated, the luminosity is 10^{13} watts. This is generated by relativistic electrons that move in the magnetic field of the nebula which has a strength of 10^{-8} to 10^{-9} tesla. Clearly the total amount of energy in the Crab Nebula is a quantity of vital interest to theorists. Most of the energy is stored in the magnetic field and in the sea of electrons, but there is no way of knowing how the relativistic energy is partitioned between the magnetic field and the electrons. Indeed it is likely that the division of energy varies throughout the nebula. However, the lowest total energy requirement occurs in an idealized egalitarian situation in which the magnetic field and the electron population have equal shares. On this assumption the total energy stores can be calculated from the synchrotron radiation theory to give a value of 10^{42} joules. This is sufficient to maintain the present luminosity for about 3,000 years, although in practice the spectrum and luminosity will change as the energy supply runs down. The total energy is equal to that radiated by our Sun in approximately 100 million years at its present luminosity.

At the highest frequencies the radiating electrons have intrinsic kinetic energies of at least 10^{14} electron volts, whereas a stationary electron has a rest mass energy of 510,000 electron volts. The great increase in energy comes about because the fastest electrons are travelling so close to the velocity of light that their mass has increased by one hundred million times. When these changed particles, called relativistic electrons, interact with the magnetic field they radiate photons with energies measured in kilo-electron-volts (X-rays) and even millions of electron volts (gamma-rays). How long can these electrons keep emitting such energetic photons? The answer is that the lifetime of the electrons responsible for the X-ray emission is only about one year in the magnetic environment of the Crab Nebula.

After a year or so the electrons are so exhausted from their whirligig round the field lines that they are only capable of sending out optical and radio photons. The enfeeblement of the X-ray electrons in a time that is a mere 0.1 per cent of the known age of the nebula proves that there must either be a means by which the electrons are accelerated again or be a machine in the nebula to replenish the supplies of electrons on a continuous basis. The same spirit of investigation shows that the magnetic field has been created since the initial explosion. It cannot possibly be the expanded relic of the original explosion because if it had been all its energy would by now have been dissipated through the overall expansion of the nebula. One of the most difficult theoretical problems associated with the Crab Nebula in the past was accounting for the magnetic field since it must have been created after the stellar explosion. As shown in the next chapter, the discovery of a very compact stellar relic, a pulsar, in the centre of the Crab Nebula gave theorists the vital clues to the origin of the relativistic particles and magnetic field.

5 · A beacon in the night

5.1 The discovery of pulsars

A totally unpredictable series of events sowed the seeds in 1964 for an astronomical sensation that led to the discovery of a unique star inside the Crab Nebula. Since the early days of radio astronomy physicists at the University of Cambridge and elsewhere had aimed to understand the powerful radio sources beyond our Galaxy. This work had contributed to the fruitful discovery of the astonishing quasars in the early 1960s. To theorists, the quasars have proved almost unyielding because they appear to generate most of their radio and optical luminosity in an exceedingly tiny volume, and yet their total emission is far greater than normal galaxies. The crucial question that demands an answer is: how can this energy be generated in the nucleus of a quasar? To gain further insight into the nature of the question in the early 1960s required higher resolution observations of the radio structures of the compact quasars, and partly for this reason high resolution telescopes were constructed at several radio observatories.

In 1964 a radio astronomer at Cambridge, Antony Hewish, found a property of the Sun that enabled him to investigate the structure of compact radio sorces without building an enormously expensive telescope. It was found that the character of the emission from a distant radio source exhibited rapid fluctuations when the Sun was close to the line of sight from Earth to the object. This effect is termed interplanetary scintillation. It occurs because there is a constant stream of matter, known as the solar wind, being blown away from the Sun's upper atmosphere. The clouds of plasma puffed out from the Sun refract the radio signals from a compact source and thus fluctuations are observed in the received signal. The effect is analogous to the twinkling of stars that you see on a clear night.

The scintillation phenomenon can be exploited to explore the tiny radio sources because the random fluctuations are most marked for the smallest emitters. As a bonus the observations also give information on the properties of the solar wind itself. Scintillation is most readily observed at comparatively long wavelengths. Antony Hewish obtained funds and proceeded to construct a cheap radio telescope at the Mullard Radio Astronomy Observatory in Cambridge. This instrument, completed in 1967, operated at the then unfashionable frequency of 81.5 MHz, at a time when most radio astronomers were striving to make high resolution observations at 1.4, 2,7 and 5 GHz. The question of observing supernova remnants or radio stars never entered into the decision to design and finance the new instrument.

Routine monitoring of the scintillations started in July 1967. For the first time in the history of their subject, radio astronomers operated a telescope of high sensitivity that had a large portion of the sky under weekly surveillance. The receivers connected to the telescope were specifically designed to be sensitive to the rapidly varying response associated with the random refraction of radio waves in the solar wind.

The Cavendish Laboratory at Cambridge, which is responsible for research in radio astronomy, has a long and honourable tradition of conducting fundamental science as cheaply as possible. This economy influenced the design of the scintillation telescope in an interesting way. Instead of the signals from the instruments disappearing directly into the maw of a small on-line computer, they were used to activate pen-recording voltmeters. Every week the pens churned their way through 130 metres of graph paper, all of which had to be carefully scanned by eye for responses from scintillation sources. This routine task was assigned to a research student, Jocelyn Bell.

Towards the end of August 1967 Jocelyn Bell noted a twinkling radio source that jerked the voltmeters into action around midnight. The detection of a scintillator at that time of night seemed an unlikely state of affairs. The Sun is then far below the horizon; therefore at midnight the line-of-sight to a radio source above the horizon makes a large angle with the Sun, where the solar wind is weak. At first it looked as if interference generated in the locality of the observatory must be responsible for the apparently spurious signals.

However on 28 November 1967 the Cambridge radio astronomers recorded the strange emission with their chart recorders working at high speed. For the first time this observation showed that the emission consisted of a series of sharp pulses. The discovery of the pulsars, objects that emit regular pulses of radio waves, dates from that time.

Further investigation showed that the radio signal consisted entirely of sharp pulses, separated in time by 1.337 seconds, and repeated with incredible accuracy. The defining property of a pulsar is that its radio emission is a highly regular sequence of sharp bleeps. At the outset it was obvious that a totally new type of celestial object had been discovered as a result of the careful investigation of an unexpected response in an instrument. How many scientists would have dismissed the initial spurious signals as 'obviously interference' and proceeded no further? If a computer had been in charge would it have been programmed to look for unwelcome or unexpected as well as anticipated responses? At times in science the boundary between a great discovery and a great disaster is narrow.

Within a few weeks the Cambridge group had found three more pulsars, and by the end of the first year various observatories had located ten of them altogether. From the beginning priority was placed on matching the pulsars to optical counterparts, but this endeavour was unrewarded. It seemed that pulsars only sent radio bleeps, not visible flashes.

Pulsars propelled a shock wave through departments of theoretical astronomy when the news of the Cambridge discoveries was first given in a seminar by Hewish and then in the columns of the leading scientific journal *Nature*. Contrived models to explain the wretched quasars, then a research area were immediately pushed to one side. Instead the remarkable pulsars came to the forefront of research. Interest centred on the clock mechanism inside a pulsar: how could it keep such regular time, beating away with an accuracy of one part in a hundred million or even better? Attention focused on the properties of the evolved white dwarf stars. A star like our Sun will become a white dwarf when it has exhausted all its energy supplies. These stars are typically a few thousand kilometres in diameter and they can vibrate radially with a cycle time of about one second. Hewish realized that a white dwarf star could be set ringing like a bell, every shiver sending out a radio bleep.

It was not long, however, before observers knew of pulsar chattering away faster than once a second. In some cases the pulse itself lasted less than one-twentieth of a second and this implicated without doubt a very small object as the driving mechanism of the pulsar clock. You can see that this is so: a beacon cannot be turned on and off once a second to give a pulse less than 50 millisecond in duration unless it is much smaller than a light second in extent. Hewish appreciated from the beginning that he was dealing with flashing stellar objects no larger than the planet Earth, and almost certainly a good deal smaller. The only stars of planetary size that had been observed were the white dwarfs, although theoretical work had been done on much smaller and more esoteric entities known as neutron stars. Methods were devised for making these white dwarfs vibrate in a way that would produce the pulses.

Within the year Large and Vaughan working in Australia with Bernard Mills at the University of Sydney made a further discovery that seriously challenged the white dwarf pulsation model. While sweeping the southern skies for new pulsars they netted a fast pulsar with a repetition frequency of only 89 milliseconds. Only the most contrived white dwarf model could shiver that often. The rapid pulsar lay in the heart of a gigantic supernova remnant in the constellation Vela. This filigree shell of cosmic wreckage is draped across five degrees of the sky around declination $-45°$. The fast pulsar is less than one degree from its centre. Mills and his colleagues considered that the association of the pulsar and the supernova remnant could not be a freak coincidence. In the announcement of their remarkable discovery they stated with laudable prescience that an identification of the pulsar with the stellar relic of a supernova seemed a reasonable hypothesis. Certainly their work marks the first firm observational link between the pulsars and the death of massive stars in a supernova explosion.

5.2 The pulsar in the Crab Nebula

Prior to the discovery of pulsars, new and important information on the radio emission from the Crab Nebula had been obtained in 1963–64, when A. Hewish and Sam Okoye found a compact source of radio emission within the nebula by means of observations at 38 MHz, carried out at Cambridge. Scintillation of the radio waves indicated that the angular dimensions of the compact part of the Crab Nebula were less than one-tenth of a second of arc. At low frequencies this diminutive component accounts for approximately one-fifth of the flux from the nebula.

After the discovery of a pulsar in the Vela remnant astronomers lost no time in looking at the Crab Nebula. The existence of the point source of radio waves stimulated them to guess that a young pulsar might lurk within the synchrotron machine. Professor T. Gold of Cornell University, New York, encouraged observers at Arecibo to look at supernova remnants. Preliminary observations at the U.S. National Radio Astronomy Observatory in West Virginia soon hinted that there were two pulsars (NP 0527 and NP 0532) in the vicinity of the Crab Nebula. Following up this suggestion, scientists at the Arecibo Observatory in Puerto Rico made the epochal discovery of a pulsar with a period of 33 milliseconds in the Crab Nebula on 9 November 1968. Even the most ardent supporters of the white dwarf mechanism had to admit defeat with observation of such a short period. Only one mechanism now seemed possible: the pulses must come from an incredibly dense neutron star, spinning around inside the nebula at thirty times per second.

Although the searches for optical counterparts of pulsars had drawn a blank, observers realized that the chances of success would be greatly increased if a search were made for flashes of light in the vicinity of pulsars. They argued that if all the light from pulsars emerged as a series of flashes, although the average intensity might be too small to register on ordinary celestial photographs, the individual flashes might be strong enough to trigger a high-speed photometer. Plans to look for pulsars in this way were soon underway at the optical Observatories in Cambridge, and at the Steward Observatory in Arizona.

Initially the group at the Steward Observatory had intended to make their observations by looking for pulses at the white dwarf oscillation frequency. However the Arecibo researchers announced the discovery of the fast pulsar after scientists at the Steward Observatory had already planned out their use of observing time. Nevertheless there was still enough time before going to the telescope on Kitt Peak mountain sixty miles away for the astronomers to change their plans so that the Crab Nebula could be scanned for rapid optical flashes. The method employed was to feed the signal from an electronic detector of light into a computer that could average and display successive cycles of a waveform in phase. In order to do this the computer has to be given an instruction on what cycle time to use in the averaging process. To search for flashes from the Crab Nebula the appropriate cycle time is the exact repetition frequency of the pulses as received on Earth at the moment of observation. Because the Earth is both rotating on its own axis and in orbital motion round the Sun the precise repetition frequency changes continuously on account of the Doppler effect, coupled with the slight change in the angle between the Earth's direction of motion and the line of sight to the pulsar. The observers at Steward, W. J. Cocke, M. J. Disney, and D. J. Taylor, duly worked out the apparent period of the Crab pulsar, fed it into the computer, and pointed the 90-cm telescope at the nebula. They waited. After four nights of good weather they had found nothing. They started to pack up.

Then news came through from the Observatory offices in Tucson that the next person scheduled to use the telescope would not after all be coming to the mountain because his wife had fallen ill. On hearing this, Cocke, Disney and Taylor began to set up the telescope in order to use the unexpected bonus of two further nights of observing time. As a true scientist, Mike Disney decided to make one last check that everything had been in order on the prevous four nights. While engaged in this check he realized that an elementary arithmetical error had been made in the calculation of the expected frequency for flashes: the correction due to the Earth's motion had been put in with the wrong arithmetical sign! The effect of this was to cause the computer to sum successive waveforms slightly out of register, a procedure guaranteed to smear out any pulses. The correct number was entered. The observers settled down as clouds closed in.

Optical astronomers hate cloudy weather, but nothing can be done except to sit and wait patiently for the sky to clear. By 9 p.m. local time conditions were looking better, and observations commenced shortly afterwards. After only ten minutes on the Crab Nebula the averaging computer was showing a definite waveform that had a pulse in it. At last astronomers had found a pulsar that flashes, after months of fruitless searches!

The discovery happened at 0330 Universal Time (Greenwich Mean Time) on 16 January 1969. It was one of the greatest events in the history of modern astronomy. The Crab Nebula had a pulsar that flashed away at the same rate in the radio and optical wavebands. Telegrams, telex messages, and telephone calls flew out in all directions to alert astronomers elsewhere. This dramatic message was telexed by the International Astronomical Union to all major observatories.

NP 0532 OBSERVED OPTICALLY JANUARY 15–16 FIVE SECONDS NORTH FOUR SECONDS EAST OF SOUTH PRECEDING STAR OF CRAB CENTRAL DOUBLE ESTIMATED ERROR FIVE SECONDS GEOCENTRIC PERIOD 33085 MILLISECONDS WIDTH FOUR MILLISECONDS OCCASIONAL SECONDARY PULSES APPROXIMATELY MIDWAY BETWEEN PRIMARIES INTEGRATED VISUAL MAGNITUDE EIGHTEEN PEAK MAGNITUDE FIFTEEN COCKE DISNEY TAYLOR STEWARD OBSERVATORY. MARSDEN

Other telescopes on Kitt Peak and the McDonald Observatory's 2.1 metre reflector swung into action. On 19 January the McDonald Observatory had confirmed the result, and other centres soon added more weight. By 3 February Joe Miller and Joe Wampler of the Lick Observatory, California had pictures which enabled the pulsar to be identified certainly with the south-west star of the pair in the centre of the nebula. This is the star that Baade and Minkowski had suspected twenty-seven years previously was the relic of the supernova. And it is the component from which the waves seem to spread out. Furthermore it coincides with the low-frequency compact radio source.

Miller and Wampler used an elegant television technique for their work. They found that the optical pulsar is at least fifty times brighter at the maximum of the pulse than at the minimum. In their experiment they fixed a television camera coupled to an image intensifier at the focus of the Lick Observatory's 3-metre reflector. In front of the camera they placed a rotatable disc with a series of six slots.

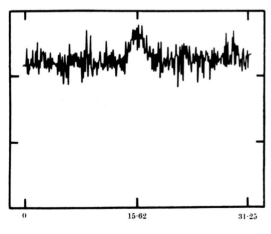

Fig. 20 First optical pulse observed at 0330 Universal Time on 1969 January 16 at the Steward Observatory, Arizona. About 5000 pulses are superimposed.

0   15·62   31·25

The slotted disc rotated at a rate that resulted in the television camera always seeing the pulsar during the same part of its cycle. With this system the south-west star in the Crab Nebula could immediately be seen as the one that turned on and off, whereas the other stellar images were of constant brightness. The images of the pulsar on conventional photographic plates therefore represent the time average of the optical brightness of the pulsar, which has not changed appreciably since the end of the nineteenth century. Although the Miller–Wampler result did not come until three weeks after the dramatic announcement from the Steward Observatory it was important because they showed that the pulsar flashed while their apparatus registered a steady signal from other stars in the field of view. This eliminated the possibility that the results from Steward and elsewhere were due to spurious signals in the electronic devices.

In February 1969 it was found that the Cambridge Observatories had detected the flashes from the Crab Nebula a full two months before Cocke, Disney, and Taylor, without being able to show that this was the case. A fast photometer had been built at the Observatories and used at the prime-focus of a 91-cm reflector that is located within a hundred metres of a brilliantly illuminated highway. (The authorities responsible for roads are no respecters of optical astronomy.) The output from this detector did not go straight to a computer, as at the Steward Observatory, but came out instead as a series of holes in a paper tape that had to be processed by a large computer in the University. On 21 November 1968 news of the radio pulsar in the Crab Nebula had reached Cambridge, U.K. as a telegram from the International Astronomical Union's clearing house in Cambridge, Massachusetts.

Roderick Willstrop scanned the central part of the Crab Nebula on 24 November 1968 and obtained eleven and a half minutes of data, consisting of 46,986 rows of holes on the paper tape. However, these data were not analysed because the University's computer was so overwhelmed with work for other scientists and students that very low priority was given to the running of jobs that required a large storage capacity. Willstrop's earlier tapes on other pulsars had given null results, and so there was a reluctance to 'waste' any more computer time. When the announcement arrived from Steward Observatory this tape was processed and the pulses confirmed. So the flashes from the Crab Nebula were first recorded in Cambridge, only 8 km from the radio telescopes that had started it all, but recognition

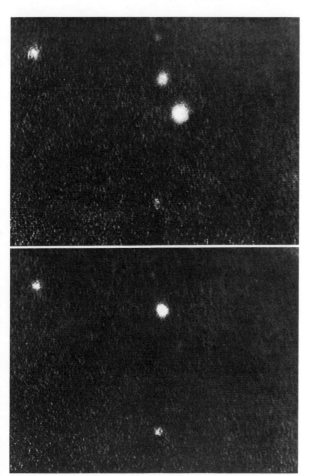

12. The stars near the centre of the Crab Nebula. This special exposure photograph from the Lick Observatory shows the pulsar caught in the act of turning on and off thirty times a second.

of the flashes must be attributed to Cocke, Disney, and Taylor. The conservative Cambridge approach that resulted in the initial discovery of pulsars failed to reward an optical astronomer so richly.

5.3 The X-ray pulsar

The Crab Nebula continued in its capacity to surprise astronomers when H. Friedman and his group at the E. O. Hulbert Center of the U.S. Naval Research Laboratory discovered pulses in the X-ray region of the spectrum, a result that was soon confirmed by other rocket and balloon flights of X-ray detectors. An experiment by a group at the Massachusetts Institute of Technology deserves special mention to illustrate the way in which different research groups started to collaborate in order to understand the Crab pulsar. Scientists at M.I.T. modified a proposed experiment in response to the discovery of optical pulses because a search for corresponding bleeps at X-ray frequencies then assumed considerable importance. In the event the N.R.L. group made a positive detection one week before the M.I.T. rocket was ready for launch.

This positive result persuaded optical astronomers at the McDonald Observatory in Texas and the Hale Observatories in California that simultaneous monitoring of the pulsations from the Crab pulsar should be carried out for one hour on 26 April 1969, at the time of the M.I.T. rocket flight. Analysis of the X-ray data immediately revealed pulses well above the background X-ray intensity of the Crab Nebula itself, and these pulses had the same waveform as the optical pulses. Comparison of the data from the Hale Observatories and McDonald Observatory with the X-ray counts showed that the pulse arrives at the Earth in both the optical and X-ray bands of the electromagnetic spectrum. Quite apart from the intrinsic importance of the result to astrophysicists, there is also a bearing on fundamental physics: the Crab pulses have traversed 6,500 light years of space, yet they arrive simultaneously at radio, optical, and X-ray wavelengths. This shows to a very high order of accuracy that the velocity of electromagnetic waves (the speed of light) does not depend upon the wave frequency.

Successive experiments from space pushed the observations of the pulsed component in the Crab Nebula to even higher energies, and the pulsar continues to register even in the hard gamma-ray region. It is detectable right up to and beyond 1,000 MeV of photon energy, and follows a somewhat flatter energy spectrum than the nebula. The pulse shape resembles that seen at lower energies, although the number of photons actually detected at high energies is too small to enable a detailed comparison to be made.

Attempts to find X-ray and gamma-ray pulses from other objects have only served to underline the importance of the pulsar in the Crab Nebula. This is the only one, out of nearly two hundred, that is readily seen at optical, X-ray and gamma-ray energies. The fact that its emission is almost entirely in the form of a pulse, of similar shape, at every observed point across fifteen decades of the electromagnetic

spectrum, sets an extremely severe problem for theoretical astronomers: how can you make a clock that ticks thirty times a second,
maintains an accuracy of one part in a hundred million, and transmits
its bleep simultaneously across every frequency from 38 MHz radio
waves right up to 1,000 MeV (3×10^{22} Hz)?

In 1977 astronomers at the Anglo-Australian Observatory, in
Australia, finally detected optical flashes from the Vela pulsar,
another of the fast pulsars. This one is fainter than the Crab pulsar.
Its average brightness is about 26 magnitude, which makes it the
faintest object ever detected from Earth. It now seems rather unlikely
that more than a handful of pulsars can ever be detected by optical
telescopes on Earth. However the operation of the space telescope
will permit stellar observations to about 29 magnitude, at which level
feeble pulsars may be detectable optically.

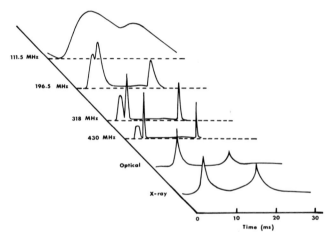

Fig. 21 The average pulse profiles of pulses from the
Crab Nebula pulsar.

5.4 The pulsar's signature

Studies of the shape of the pulse from the Crab Nebula have shown
that it is unique. The pulse shape is determined by averaging the data
for many thousands of pulses. This procedure shows that the average
pulse profile at a radio frequency of 430 MHz consists of an intense
but brief main component that lasts for about 250 microseconds,
which is followed 13.37 milliseconds later by a broader and weaker

component termed the 'interpulse'; the interpulse does not occur exactly between successive main pulses. Preceding the main pulse by 1.6 milliseconds there is another broad and weak component that is known as the 'precursor'.

The radio signature of the pulsar changes markedly at lower frequencies, and in this respect the Crab Nebula pulsar is unlike most other galactic pulsars. At lower frequencies the pulse shape becomes more washed out as the well-defined features seen at 430 MHz become broadened. These changes are due to the effects of radio wave propagation in the interstellar medium between ourselves and the pulsar. What happens is that individual waves from the pulsar get scattered by the electrons in the interstellar medium, and thus the signal that actually reaches our radio telescopes is the superposition of signals from very many slightly different paths through the medium. This effect broadens out the waveform in a way that depends strongly on frequency such that it is most noticeable at long wavelengths. At a sufficiently low frequency the waveform will be smeared so much as to be indistinguishable from continuous steady emission. The radio spectra of the pulsar and the low frequency compact radio source match each other smoothly, suggesting that the two objects are in reality but one.

A striking characteristic of the Crab pulsar is the occasional emission of jumbo pulses at random intervals. Something like one pulse in ten thousand is very intense. In fact the strongest pulses instantaneously exceed the total radio emission from the Crab Nebula itself by an order of magnitude. During these fleeting moments the pulsar is the brightest object in the sky at metre wavelengths. It is the main pulse itself which features the strong amplification. Since the effect is not shown by the interpulse or by other pulsars it cannot be due to chance enhancement of the signal by fortuitous scattering in the interstellar medium but must be intrinsic to the pulsar.

The discovery of the strong pulses led to the resolution of a problem concerned with the initial findings, in 1968, of two pulsars near the Crab Nebula. D. H. Staelin and E. C. Reifenstein had observed strong brief pulses from the vicinity of the nebula in 1968, and they hypothesized that there were either two pulsars, both of them highly sporadic, or some new type of pulse-emitting object. Subsequently the famous Crab pulsar was found at Arecibo, and the other pulsar, located nearly one and a half degrees from the nebula, was found to have the longest known period: 3.75 seconds. Staelin and Reifenstein used the 90-metre radio telescope at the U.S. National Radio Astronomy Observatory; the two pulsars were therefore named NP 0531 (Crab Nebula pulsar) and NP 0527, the N signifying NRAO. The four numerals give the right ascension, which was initially poorly determined; the object in the Crab Nebula is now called NP 0532.

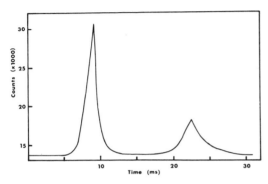

Fig. 22 The light curve of the Crab Nebula pulsar, showing the needle-sharp peak in the main pulse.

In the optical domain the pulse waveform is much less structured than in the radio region. Both the main pulse and the interpulse are slightly asymmetrical: in the main pulse the signal falls off more sharply than it rises, whereas the opposite situation prevails for the interpulse. The peak of the main pulse is surprisingly sharp, and it plummets down in only 40 microseconds. The needle-sharp peak has not yet been resolved. The X-ray signature is interesting because both the pulse and interpulse have about the same intensity. At all frequencies the signal between the pulse and interpulse, a time interval of 13.5 milliseconds, never returns to the zero level observed just before the principal pulse registers.

5.5 Polarized pulses

The strong radio pulses are typically twenty-five per cent linearly polarized and they exhibit slight circular polarization as well. No general rule governs the polarization properties of individual jumbo pulses; sometimes they are strongly linear, and at other times circular. On the other hand, the average pulse polarization, summed for thousands of individual pulses, shows systematic trends. The precursor pulse is completely linearly polarized, which tells us that the emission must come from a small and highly uniform region of magnetic field that is presumably near to the pulsar. The linear polarization of the main pulse is twenty per cent and of the interpulse thirteen per cent. A few pulsars show rotation of the plane of polarization while a pulse is being received; this is very marked for the Vela pulsar, where the plane swings through nearly 180° during each pulse. The Crab Nebula pulsar takes no part in this type of behaviour. Observations of the plane of polarization made at several frequencies have clearly shown that the plane of polarization is rotated by a constant amount through the pulse as the electromagnetic waves travel through the interstellar medium beyond the nebula. This is, of course, a feature exhibited by all radio waves as they journey through the patches of electrons and magnetic field in the Milky Way, and is merely a manifestation of the Faraday effect. The polarization measurements for the Crab Nebula pulsar show that there is no medium capable of causing rotation within the immediate vicinity of the pulsar, and therefore there is no significant concentration of thermal electrons close to the pulsar.

5.6 Falling behind schedule in interstellar space

In addition to the rotation of the linear polarization, the electrons in space have a further subtle effect on the progress of electromagnetic radiation. The electric field in the wave jiggles the free electron encountered in space. The motion of the electrons sets up, effectively, another electromagnetic wave in competition with the original wave. From the experimental point of view the net effect is that the velocity of the waves received by a telescope is slightly frequency dependent. We are familiar with the refraction of light by a glass prism, which breaks or disperses the light into its constituent colours. Essentially the interstellar medium with its attendant electrons acts as a refracting medium for the radio waves. The refractive index depends on the frequency of the wave. For a uniform signal this has no noteworthy consequences. If, however, the signal is composed of a series of pulses, then the arrival times suffer a delay that is frequency dependent. Lower frequencies get more behind schedule than high frequency waves as they battle through the electrons and fields. Arrival times of the blips from pulsars show this delay, which is known as dispersion. The effect is important astrophysically because the amount of delay provides a measure of the electron density along the line of sight that is independent of the magnetic field strength. The Faraday rotation, on the other hand, depends on both electron density and magnetic field. Therefore, by measuring both Faraday rotation and dispersion for a given pulsar's pulses it is in principle possible to separate the effects of the average electron density and the average magnetic field along the line of sight.

For the Crab Nebula pulsar the observed dispersion indicates that on the average the electron density along the line of sight is 25,000 electrons per cubic metre. This is substantially fewer than was thought to be the case generally before pulsars were discovered: estimates of 10^5–10^6 electrons per cubic metre for the density of interstellar electrons were commonplace. The dispersion measure of the Crab pulsar is roughly what we would expect along a typical line of sight through the Galaxy for a distance of 2 kpc. A simple analysis of the rotation gives a value of 10 microgauss (10^{-9} tesla) for the average interstellar magnetic field along the line of sight; the analysis tells us nothing about the value of the field at right angles to the line of sight. However, as in the case of the electron density, the deduced magnetic field agrees with the value expected from independent studies. So, even if the Crab Nebula itself is rather extraordinary we can take comfort in the fact that we are carrying out our observations of it along an average path through the Milky Way that is not affecting the radiation in an unknown or unpredictable fashion.

Observations over several years have shown no overall change in the amount of dispersion in the pulses. However, short term (of order months) variations do occur. Quite likely these small changes are caused by clouds of electrons that drift through the nebula and cross the line of sight.

5.7 The clock runs slow

Pulsars are magnificent natural clocks, and the Crab Nebula pulsar is, in many ways, the most impressive of them all. The time of arrival of a given pulse can be measured to an accuracy of about 30 microseconds, and can, if desired, be compared to the very best laboratory standards, which are atomic clocks. The accuracy of 30 microseconds from a single observation can be much improved on if pulses are counted and timed, even intermittently, for, say, a year, then an accuracy of one part in 10^{12} can be attained in the measurement.

All this assumes, of course, that astronomers do not miscount the pulses. In a year the Crab pulsar makes one thousand million bleeps. Naturally it is impossible to observe the pulses when the Crab Nebula is below the local horizon, but that does not matter. The internal clock is sufficiently accurate that allowance can be made for the pulses not detected between one observing session and the next. In a similar way, for example, we know that an ordinary clock will tick 36,000 times in ten hours without having to count all the ticks or measure each one individually. In the case of NP 0532 the internal timing mechanism is phenomenally accurate, and thus its pulsing behaviour is now tied down to a few parts in 10^{13}. This is comparable with the accuracy reached in fundamental standard clocks, which have an inherent stability of a few parts in 10^{13} also.

Timing observations of NP 0532 reveal a considerable amount about its physical properties even though these discoveries do not come up to the early hopes of astronomers. Unfortunately the super-accurate period is disturbed by a poorly-understood random process that starts to corrupt the data at the level where really fundamental experiments could be carried out. The noisy component in the period behaviour means, for example, that the pulsar in the Crab Nebula is not superior to the best man-made clocks as a fundamental standard for physicists. This is a pity since the attraction of a standard clock in the sky, that all physicists working the northern hemisphere could observe, is considerable.

The method of deducing the time of arrival of the pulses is interesting. Let's look at the problem in the following way. The Crab Nebula is far beyond the solar system and it contains a natural clock that rivals the best clocks on Earth. You want to compare this natural clock with a real clock, but you cannot bring the two together. Furthermore, your observations are to be conducted at different times in several observatories, while the Earth itself is rotating, and travelling around the Sun. What this means is that all the raw data actually gathered at the telescope have to be reduced to a standard form that takes account of the varying locations of the observatories and of the Earth's own varying motion through space.

The timing of the pulses themselves has to be relative to a standard time system that is transmitted continuously by radio. In order to reduce these data to the time registered by the standard, a correction has to be made to take account of the location of the observatory relative to the laboratory that contains the reference clock. An uncertainty of as little as 100 metres in the location of an observatory puts an error of 0.3 microseconds into the measurements—the light travel time for 100 metres. Positions on the surface of the Earth are typically uncertain to this degree even at an astronomical observatory because random fluctuations in the rotation rate of the Earth mean that latitude and longitude cannot be found to an arbitrarily high accuracy.

Once the arrival times of the pulses at the telescope have been established as well as possible, they must be converted to the values that would have been obtained at the centre of gravity or barycentre of the solar system; this is the most convenient 'stationary' reference point in the Earth's vicinity. To make the conversion the motion of Earth round the Sun has to be known to a high order of accuracy, and indeed perturbations on the orbit by all the planets must be taken into account. Furthermore, even though our planet is circling the Sun at only a tiny fraction of the speed of light (0.01 per cent), the theory of relativity rather than merely classical physics must be employed in order to get the desired very high accuracy in the final data. Transferring all the observations to the centre of the solar system removed the complicating factors that are introduced by the Doppler effect: since Earth is moving in a continuously changing direction relative to the pulsar, the Doppler correction to the pulse arrival times is changing continuously as well. Once you've got all these complexities under control you're set up to keep watch on the Crab clock.

The first timing measurements were made by radio astronomers, who soon discovered that the clock governing the emission of pulses is slowing down at the uniform rate of about 36.5 nanoseconds (a nanosecond is a thousand-millionth of a second) per day. This is observed as a steady increase in the period of the pulsar, which in a year lengthens by approximately ten microseconds; the actual period of the pulses is approximately 33.19 milliseconds (in 1978). Within only a few months it became clear that the pulsar slow-down rate is very regular indeed.

Researchers at Princeton University, New Jersey made the first systematic timing observations of the Crab Nebula pulsar using equipment attached to a 92-cm telescope at the University's observatory. Five years were spent accumulating the data, which were analysed by Edward Groth. In all arrival times were logged on 348 nights. Groth used orbital information for the solar system prepared by scientists at the Massachusetts Institute of Technology and a master clock at the United States Naval Observatory to reduce his data to the standard form. To extract the maximum information from the arrival time measurements, which had been made at irregular intervals, he had to develop a new technique of analysis.

From the precise observations made by the Princeton group Groth computed the period, rate of change of the period, and rate of change of the rate of change of the period! In addition to revealing these three quantities, the first two of which are known now to very high precision, the data also showed the restless or random behaviour of the Crab Nebula pulses mentioned earlier. The random background is revealed in the following way. If the data for a few years are fitted by an equation that takes into account the slow-down rate and the rate of change of the slow-down, it is found that, after all the errors have been taken care of, there are slow drifts in the arrival time. The easiest way to describe these is to say that for months on end the pulses seem to arrive tens of milliseconds early or late, slowly and unpredictably drifting between the two. This noise process seems to be completely random, and it is the fundamental limitation on the accuracy to which parameters connected with the timing observations can be derived.

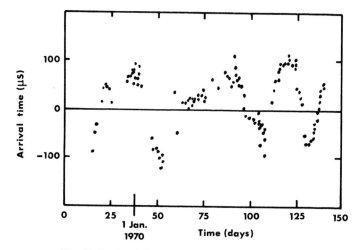

Fig. 23 Restless behaviour of the pulsar is evident in this plot of discrepancies in the actual pulse arrival time against expected arrival time. These data are from the Arecibo Radio Observatory.

Most pulsars exhibit the slowing down phenomenon, including the other fast pulsar in the Vela supernova remnant, but the Crab Nebula pulsar shows by far the fastest rate of change. No pulsars are speeding up, from which we safely infer that the systematic increase in the periods is indicative of a dwindling energy supply to the basic clock mechanism. For pulsars in which the period, and its rate of change, are both known, we can calculate a crude pulsar 'age'. For example, suppose that NP 0532 initially had a period of zero seconds; then from the presently observed slow-down rate we can infer an 'age' of 3,300 years—the time needed to lengthen from 0 to 33 milliseconds at the same slow-down rate throughout. Clearly this is too rough and ready a method of deriving an age, but it is encouraging that it comes within a factor three of the known age of the supernova remnant, confirming for us yet again the association of a diversity of Crab Nebula phenomena. An improved estimate, taking into consideration also the observed rate of change in the deceleration, would yield an age much closer to one thousand years, and an initial period about half that currently observed, i.e. 16 milliseconds or so. When the calculation is carried out for the other pulsars, an age of 12,000 years results for the Vela pulsar, and ages of millions of years for the remainder. Even though the estimates have considerable assumptions built into them, they nevertheless indicate that the Crab Nebula pulsar is unusually youthful in comparison to all the others that we can study, once again underlining its importance to astronomy.

There is a parameter of particular interest to theoretical astronomers that can be deduced from the timing data. This is a number called the 'braking index', which is a quantity that specifies the relative effectiveness of the mechanism that is slowing down the pulsars. It is found by forcing the data on the pulse frequency (i.e. pulses per second) and rate of change of this frequency to fit a power-law, such that the rate of change is given by the frequency raised to a power known as the braking index. If we denote the frequency by f and its rate of change by df/dt, then the braking index n is given by $df/dt = kf^n$, where k is a constant. The precise mathematical details need not trouble us. The points to bear in mind are that the larger the value of n derived from observations, the fiercer the pulsar brakes are being applied, and that n is a number that theorists can predict from their models. So we shall need the braking index when we come to confront theoretical playthings with the real world of the pulsars! For

NP 0532 the braking index is 2.6; this is in general agreement with the value for many other pulsars, indicating that the same types of brakes apply in pulsars of all ages.

By the end of 1969 astronomers had discovered that the two fastest pulsars unexpectedly made a sudden effort to speed up again, before resuming the slowing down pattern once more. These discontinuous jumps in the period are called 'glitches', a term derived from a Yiddish word and adopted by electronics and space engineers; you are not likely to find it in any but the most recent dictionaries. The Vela pulsar gave the first observed glitch somewhere between late February and early March 1969. Astronomers in Australia and the United States had been watching its orderly progress, when the period suddenly jumped up by 0.2 microseconds before settling back to the regular routine a few days later. The glitch enabled the pulsar to regain the pulse rate that it had had about twenty days previously. It took a year for all traces of the sudden jump to vanish from the timing data.

On September 1969 the Crab Nebula pulsar duly obliged with a major glitch, causing a speed up in the pulse frequency by about one part in 10^8. Other sudden jumps have been noted subsequently, but

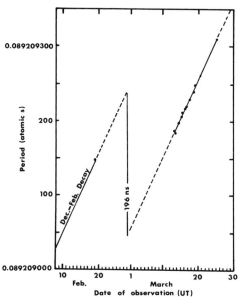

Fig. 24 The sudden step in period (glitch) of the Vela pulsar. At some time between 19 February and 13 March 1969 the period decreased suddenly by 196 nanoseconds.

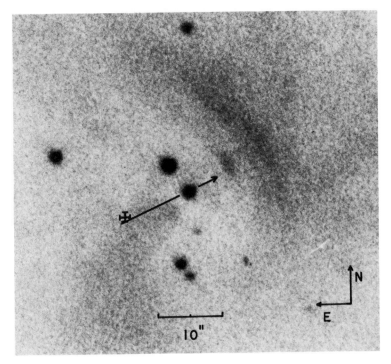

13. The centre of activity in the Crab Nebula. The motion of the pulsar since AD 1054 is indicated. Note the faint wisps ahead of the direction of motion. (Courtesy Royal Greenwich Observatory, S. Wyckoff and P. Wehinger)

the 1969 glitch is among the largest that have been reported. After the big glitch the pulsar recovered its normal behaviour on a timescale of about a week, very much less than the relaxation time of a year associated with Vela pulsar glitches. It is possible that the underlying jittery behaviour in the period of the Crab Nebula pulsar is due to a random succession of microglitches.

We see that all the timing observations give a picture of an extremely interesting natural clock inside pulsars. Not only is the basic repetition frequency held remarkably accurately, but the slow-down rate is as well. This means that the pulse period can be predicted very accurately into the future. In fact, when a timing experiment started on Britain's 2.5-m Isaac Newton telescope in 1977, times of arrival and rates of change determined three years previously had to be used to predict the expected time of arrival and the precise period.

Observations started at 6 p.m. one evening, and by 10 p.m. the pulsar was right on schedule, a remarkable achievement for both the pulsar, and for astronomers, after a gap of three years and the emission of billions of bleeps.

5.8 The pulsar's on the move

Susan Wyckoff and Andrew Murray, of Britain's Royal Greenwich Observatory, secured vital material linking the pulsar with the A.D. 1054 explosion. In 1977 they presented an analysis of plates of the nebula across the range 1899 to 1976. Over this time span of seventy-seven years they logged the pulsar's position relative to distant stars in the same field. This enabled them to deduce that the pulsar is speeding along at about 125 km per second—about four times faster than the general motion of the stars. Beyond doubt this pulsar is a high-velocity object.

An intriguing fact is that the pulsar is ploughing through the nebular gas towards the direction of the most intense wisp structures. Furthermore it is heading towards the centre of the X-ray emission. This suggests that the pulsar really is stirring up trouble: a shock wave forms where high-speed particles from the dense spinning star crash into the surrounding nebular gas.

The crucial link with the A.D. 1054 explosion is provided by retracing the motions of the filaments and the pulsar. Wyckoff and Murray find that these points of divergence coincide to within the observational errors.

5.9 Confrontation of observation and theory

This chapter deliberately concentrates on a description of the observed phenomena, keeping the theoretical models for a separate presentation. This prevents the 'contamination' of facts by hypotheses, and reduces the chance of presenting the data in a way that is model-dependent. In this section the salient points are marshalled together in order to see what can be fed into the theories.

The pulsar in the Crab Nebula flashes thirty times a second, and is so faithful that the arrival time of flashes many months ahead can be predicted with considerable accuracy. It occasionally has glitches that suddenly speed up the pulsar by a few parts per billion, but it only takes a few days to get back on schedule after a glitch. The pulse itself can be seen at radio, optical, X-ray and gamma-ray frequencies. In the radio region the pulse is delayed and its polarization rotated by electrons and magnetic fields along the line of sight. The optical and X-ray flashes arrive simultaneously; this places a stringent limit on the variation (if any) of the speed of light with frequency, and proves that the optical and X-ray arise in the same volume of space. Independent lines of thought yield an age of one thousand years for the object. It is clearly a remnant of the explosion of a star.

From this summary we can identify three principal tasks for theoretical astronomy: (1) to describe the nature of a pulsar, (2) to explain the clock mechanism, and (3) to give an account of the production of the pulses. The next chapter examines the progress in these three fields. Although the same questions apply to all pulsars, the investigation of the Crab Nebula pulsar indisputably constrains the models and shows the way forward more effectively than all the other pulsars put together.

6 · The strange world of a neutron star

6.1 A triumph for theoretical physics

In 1932 scientists at the Cavendish Laboratory, Cambridge, discovered a new fundamental constituent of matter: the neutron, an elementary particle that carries no electrical charge. It is slightly heavier (0.15 per cent) than the proton, which has a positive electrical charge. Protons and neutrons are respectively 1,836 and 1,839 times the mass of an electron. They are the basic constituents of the dense cores, termed nuclei, of the atoms. The hydrogen atom has a nucleus consisting of one proton and the helium nucleus normally consists of two protons and two neutrons, for example. In a neutral atom electrons orbit the positively charged nucleus in number equal to the proton content; so hydrogen atoms have one electron and helium atoms two. Roughly speaking the neutrons and protons have radii of 10^{-15} metres, while atoms themselves have radii of 10^{-10} metres. You can see, therefore, that almost all the mass of an atom resides in the nuclear core. The density of nuclear matter is 2.8×10^{17} kg per cubic metre, or, put more graphically, a small spoonful of neutrons or protons would weigh one billion tonnes at the surface of the Earth! Looked at in this unconventional way, the density of 'ordinary' matter is seen to be entirely a consequence of the wide spacing of nuclei caused by the roomy electronic structure of the atoms. If the atomic structure of matter could be destroyed through compression a much denser state would result. In ordinary matter, however, electromagnetic forces completely dominate the other forces known to physicists, namely the two nuclear forces and the gravitational force. But there are circumstances in which matter can collapse down to nuclear densities, as we shall see.

On the very day that news of the discovery of the neutron reached Copenhagen, the Soviet physicist Lev Landau discussed possible implications with theorists. According to anecdotal records it was Landau who first suggested that there might exist cold dense stars composed primarily of neutrons. Such exotic stars would be tiny objects having the density of nuclear matter. Quite independently, apparently, Walter Baade and Fritz Zwicky proposed in 1934 the idea of neutron stars. Furthermore they advanced the novel suggestion that these objects might be formed in supernova explosions. Zwicky, in 1939, suggested that the vast energy of the explosion probably derived from the creation of a tiny neutron star, and he is credited with the amazing speculation that there ought to be a neutron star in the Crab Nebula.

We see that from decades before the discovery of pulsars, theorists had proposed the existence of dense neutron stars, and they even linked them with the endpoints of stellar evolution. Until the finding of a rapidly pulsing radio source in the Crab Nebula, the neutron star was doomed to remain an imaginative plaything of the theorists. As explained in the previous chapter, the first few pulsars encouraged the development of models of radially-pulsating white dwarf stars. However, their size is akin to that of the planets. With the discovery of fast pulsars in the Vela and Crab Nebula supernova remnants it became obvious that entities much smaller than planets must be triggering the stream of pulses.

In a 1968 paper that preceded the discovery of fast pulsars, Thomas Gold of Cornell University championed the view that neutron stars are responsible for the pulsed emission. He argued that these must be rotating because a pulsating neutron star could not vibrate as slowly as would be required to account for the pulsar with the most ponderous pulse. Hence a picture emerged of a rotating neutron star emitting an intense beam of radiation, rather like a lighthouse. In the final paragraph of this seminal paper Gold predicted that observers would find a slight, but steady, slowing down of the repetition frequency as the spinning star gradually shed its rotational energy. Just this effect was found soon afterwards for the pulsar in the Crab Nebula. Subsequently Gold rounded off this brilliant detective work by showing how the observations of NP 0532 provided four crucial pieces of compelling evidence in favour of his rotating neutron star hypothesis: (1) theoretical work had linked neutron stars to supernovae; (2) two supernova remnants shielded pulsars: (3) only rotating neutron stars gave a precise clock mechanism; and (4) the predicted slow-down was subsequently confirmed.

6.2 The theory of collapsed stars

To put the identification of the Crab Nebula pulsar with a neutron star into its historical and scientific context, it is appropriate first to review what astronomers believe to be the final stages in the life cycle of a star. The first successful attempts at showing how a star works were made when astrophysicists constructed models of the internal structure of stars based on the idea that gas pressure supported the layers of the star against the internal gravitational forces. This idea is essentially correct.

In the early part of this century certain puzzling stars refused to fit the hot gas models. They had luminosities which were far smaller than those measured for the majority of stars with similar colours. Eventually masses were found for a couple of the stars, and the big surprise was that these were not unlike that of our Sun. What could these very dim objects as massive as the Sun be? The astounding characteristic was their density, which exceeded by many millions of times that of stars like the Sun and of all matter ever studied on Earth. These oddities, termed white dwarfs on account of their colour and small size, remained a profound mystery to astronomers until physicists had dramatically modified the entire classical view of the natural world, in the greatest revolution in thinking since the times of Galileo and Newton.

In the 1920s a host of distinguished theoretical physicists, such as Einstein, Fermi, Dirac, and Heisenberg, had shown that the world of atomic particles cannot be treated mathematically using only the concepts of classical physics.

Classical physics treats the material world in terms of the electromagnetic theory of Clerk Maxwell, the mechanics of Newton, and ordinary common sense. These 'obvious' theories all failed to give a correct picture of the behaviour of radiation and matter under *all* circumstances. For example they predicted that the orbiting electron would crash straight into the central proton. Instead a new theory, that of quantum mechanics, was needed to deal with the behaviour of the denizens of the atomic world: photons, atoms, electrons, and so forth. Everyone who encounters the predictions of quantum mechanics for the first time finds them very puzzling; the same is often true of Einstein's theory of relativity. This is because both theories deal with processes that are not encountered routinely on the level that we deal with in everyday experience. Quantum mechanics and general relativity are not needed in everyday life. Consequently analogies can be unhelpful, even misleading, and the rules governing the behaviour of elementary particles can appear to be quite arbitrary. 'Common sense' is definitely detrimental to an understanding of the atomic world!

Wolfgang Pauli elucidated a famous 'exclusion principle' that governs the behaviour of some of the atomic particles. In general terms this principle states that in the atomic world two identical particles (e.g. two electrons orbiting an atom or two neutrons in a nucleus) may not have identical energy states. If we consider a definite volume of space with some electrons in it, the Pauli exclusion principle tells us that we may not put extra electrons with any energy we please into the volume. The exclusion principle teaches that electrons can only be added if they have certain energies. Any addition will only be possible if the new electrons have energy states which differ from those already present. Fortunately the spacing of energy levels is normally exceedingly fine, so a multitude of empty states is generally available to newcomers. If a volume is decreased, however, the number of energy levels inside that volume decreases also. Once the energy levels available in a volume start to fill up, the constraints on how particles can move begin to introduce unusual physical effects. In fact, as we shall see, the exclusion principle is at work inside collapsed stars, sustaining them against the crush of gravity.

Ordinary stars can support themselves against the inward pull of gravity through gas pressure because the continuous generation of energy in the central core makes the stars hot, and a hot gas exerts a pressure. When the fuel in the core of a star is burnt out much of it consists of heavy elements, up to and including iron. Once a nuclear fuel is exhausted, as it is when nuclei in the neighbourhood of iron have been synthesized in the stellar interior, the star starts to go cold. A cold star must inevitably shrink under the force of its own gravity; as it shrinks the internal gravitational force intensifies because the star's mass is concentrated into a smaller volume, crushing the material yet tighter. Temporary re-arrangements of the structure of a star might possibly lead to limited new sources of energy. For example, in some cases significant hydrogen may remain in the stellar envelope, and it may be possible to stave off collapse by processing this. But eventually all the fuel that can be used has been used and the star has no option but to cool, and, inevitably, to shrink.

The internal pressures in a shrinking star are sufficiently high that all semblance of atomic structure is soon wiped out. We can legitimately view a collapsing star to be composed of nuclei (principally iron) embedded in a sea of electrons. Thus we have a situation in which a population of electrons is being constrained to move in an ever-decreasing volume as gravity shrinks the star. Under certain circumstances the Pauli exclusion principle is able to rescue the star from indefinite collapse. The main condition is that the mass of the shrinking star must be less than 1.4 solar masses. The population of electrons, which behaves in some respects as a gas, finds fewer and fewer energy states available with the decreasing stellar volume. The lowest energy states get filled up first, and when most of the available levels are full the electrons exert a pressure known as degeneracy pressure.

This degeneracy pressure has no counterpart in the classical world. It arises quite simply because the electrons are jammed into all the available energy slots. No further shrinkage can occur because this would reduce the number of energy levels to fewer than the number of electrons which is impossible in the quantum world. Inside a cold star the electron degeneracy pressure can stabilize the matter against the inward pull of gravity provided the mass is less than 1.4 solar masses. The white dwarfs are therefore degenerate stars supported entirely by electron pressure. Perhaps we should call them 'electron stars'?

When the neutron was discovered it was apparent that it might be possible to have stars supported by the pressure of a degenerate neutron gas, since neutrons show the same statistical behaviour as electrons in the microscopic world. It appears to have been George Gamow who first specifically derived, in 1937, some of the properties of stars held up by neutron pressure. But before looking into this we ask how can stars made mainly of neutrons be made in the first place?

Imagine a star being squeezed even more tightly than a planet-sized white dwarf. This might happen if the star exceeds the maximum allowed mass for a white dwarf, or if stupendous explosions in its outer layers send strongly compressive shock waves down through to the core. As the central density of such an object rises a point is always reached at which it is energetically favourable for the electrons to be driven inside the massive nuclear particles. Recall that the nuclei consist of protons and neutrons. Essentially what happens is that the protons and electrons combine. They cancel their electrical charges in this union and thus produce (neutral) neutrons. Hence the nuclei inside a star with sufficiently high central density become progressively neutron-rich. Contrast this behaviour inside the dense core with what happens in ordinary matter. The free neutron—one far from any other neutrons or protons—decays into a proton and an electron on a timescale of twelve minutes, because the proton-electron combination has less energy than a neutron. Similarly, in ordinary matter the number of neutrons in a nucleus is seldom more than one and one-half times the number of protons and is often less than this. Any excess neutrons that happened to be present in the nucleus would soon convert to protons, with the consequential emission of electrons (one aspect of radioactivity). This state of affairs cannot exist inside the neutron star however because the available electron energy states are full. So, once a neutron has formed it *cannot* decay into a proton and an electron. The Pauli exclusion principle blocks the decay because the electron which would be created cannot exit into a vacant energy level. Hence the star has no means of disposing of any excess neutrons, which accumulate through the union of protons and electrons. The net result is that the core of the neutron star gets progressively richer and richer in neutrons, at the expense of protons and electrons. These

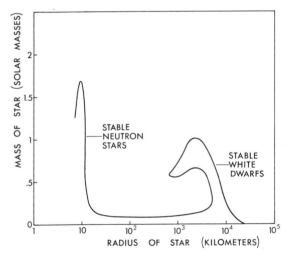

Fig. 25 Mass-radius graph for high-density stars shows two possible stability regions that correspond to the two classes of condensed stars: white dwarfs and neutron stars.

neutrons are able to arrest the gravitational collapse of the star because they exert a neutron pressure due to the exclusion principle, just as the electrons exert a pressure in white dwarfs.

Theorists have calculated that the range of allowed masses for neutron stars is probably between 0.2 and 1.4 solar masses; there is uncertainty about the upper limit because physicists do not know enough about the precise way in which neutrons interact with each other at the very high densities involved. In fact the average densities of neutron stars span a narrow range of 10^{17}–10^{19} kg/cubic metre. This is roughly one million times denser than a white dwarf and implies, for a spherical star, a radius in the region of 15 km.

The range of neutron star masses parallels that of the white dwarfs. It is likely that stars reach the white dwarf configuration relatively harmlessly, and, once they are white dwarfs, there is no incentive for them to condense to the more extreme configuration. The electron pressure can hold the star in equilibrium against gravity indefinitely. White dwarfs are therefore an important endpoint to the possible life cycle of stars. Neutron stars are probably formed in violent events, when the collapse proceeds so rapidly that the electrons simply do not have time to reverse the inward rush of material. Hence the

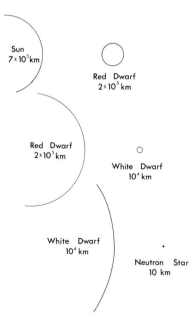

Sun
7×10^5 km

Red Dwarf
2×10^5 km

Red Dwarf
2×10^5 km

White Dwarf
10^4 km

White Dwarf
10^4 km

Neutron Star
10 km

Fig. 26 Relative sizes of various stars compared. The sizes, from Sun to neutron star, encompass a large range, which is shown here as three pairs: sun—red dwarf; red dwarf—white dwarf; white dwarf—neutron star. Even so the neutron star is exaggerated ten times!

association of a known supernova remnant—the Crab Nebula—with a known neutron star—NP 0532—is important in demonstrating the internal consistency of current theoretical work on the death of stars.

6.3 The surface of a neutron star

The discovery of the Crab Nebula pulsar delivered new impetus to the study of neutron stars. Many of the results are highly speculative and extremely interesting excursions into a world quite unlike any ever explored before by physicists. It might seem extraordinarily presumptive, for example, to try and explain what the surface of a neutron star is like. The fact that physicists can give us a rough picture of the surface reflects the close interplay between fundamental physics and the science of astronomy. Astronomers have found a class of object that adds reality to all those experiments inside expensive particle accelerators. Neutron stars are macroscopic objects in which we can see microphysics in action.

Suppose you could stand on a neutron star; how much would you weigh? You would be squashed into an extremely thin layer because gravity is 10^{11} times stronger at the surface of the Crab Nebula pulsar than at the surface of the Earth. This book would weigh over 25 million tonnes, and you would tip the scales at a few billion tonnes. If you fell a distance of one metre from rest you would hit the ground at a speed of 1,000 km per second (two million miles an hour). Furthermore, the rate of change of gravitational force is so great (0.01 per cent per metre) that tidal forces would tear you apart! A neutron star is such a dangerous object that only theorists and computers can ever survey its crust.

A neutron star probably has a layered structure. The outer parts, where the pressure is lower, have different properties to the inner core which is being crushed by the weight of the overlying matter. The outer layer is a solid crust. At the surface itself the neutrons and residual protons form heavy atomic nuclei that lock together into a lattice that is permeated by a sea of free electrons. The conditions inside an ordinary metal are broadly similar to this, except that the atomic nuclei retain some of the orbital electrons. In the neutron star's metallic lattice most of the nuclei will be iron since this element is at the end of a chain of reactions that fuel stars. However, deeper inside the crust, the higher pressures cause more massive nuclei to form. Among the heaviest nuclei that are stable in the crust are molybdenum-124 and krypton-118. These nuclei contain eighty-two neutrons, which are able to arrange themselves into stable closed shells inside the nucleus. Neither nucleus can be studied in a terrestrial laboratory because at normal pressures they immediately decay by emitting electrons. A slice through the crust of a neutron star would show a series of metallic layers, with the heaviest nuclei at the base of the crust.

Once a mass density of 4.3×10^{14} kg per cubic metre is reached, at a depth of about a kilometre, an intriguing phenomenon known as neutron drip occurs. Nuclei, such as krypton-118, are so rich in neutrons that any further increase in the pressure causes neutrons to ooze out of the nuclei. Neutron energy states in the lattice start to fill up as a sea of neutrons forms around the nuclei. When the density of nuclear matter is reached, at a value around 5×10^{16} kg per cubic metre, the nuclei dissolve into an ocean of neutrons, protons and electrons. The 'billiard balls' of classical atoms have been put through a liquidizer and turned into a quantum ocean. It is thought that 99.5 per cent of this fluid is made up of the neutrons. In this esoteric liquid the neutron degeneracy pressure is sufficient to stabilize the star against collapse, and to support the crust, which is 1 km or so thick. Within the crust itself, particularly the upper layers, the free electrons are the main source of pressure.

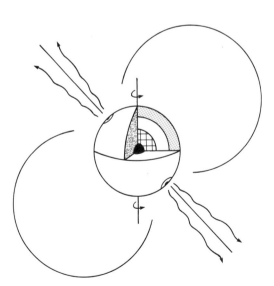

Fig. 27 Postulated interior structure of a neutron star. A hyperon core (solid black) is surrounded by a neutron crystal lattice (squares). Then comes neutron superfluid (unshaded) and the crust (dotted). The magnetic field and the beaming of radiation along the axis are also indicated.

6.4 Superfluid interiors of neutron stars

Theorists consider that beneath the crust of a neutron star, like the one in the Crab Nebula, the material will behave in many respects as a fluid. However, we must remember that the behaviour of this fluid is governed by quantum mechanics not classical hydrodynamics. Hence the liquid interior has some surprising properties not shared by normal liquids. It is superconducting and is a superfluid. These states of supermatter had already been studied by laboratory physicists before the discovery of neutron stars, but had never previously been encountered in celestial objects.

The superfluidity inside a neutron star arises in the following extraordinary way. At the extreme densities of matter that we are now concerned with there is a small attractive force between neutrons. The result is that the neutrons can form themselves into pairs of neutrons, and these pairs can be considered crudely as individual particles. Now a bound neutron pair does not follow the same behaviour as an independent neutron. In particular the bound pair is not subject to the restrictions of the exclusion principle. Hence the bound pairs of neutrons can all be in the same energy state even though this is impossible if they choose to act as individuals. This happy state of affairs means that the fluid of neutron pairs will offer no resistance to motion. It has no viscosity: in the ordinary flow of a liquid energy is expended by driving the particles of the fluid into a range of energy states, and this gives the fluid a resistance to flow because work must be done to move it. Now in a superfluid all the particles are in the lowest state, and they have no incentive to redistribute their energy in a less egalitarian way. Hence the fluid flows without restriction.

In a similar way the residual protons in the liquid interior can form pairs. But these pairs carry the two positive electrical charges of the protons. As in the case of superfluidity there is no resistance to the motion of these charged pairs, and the proton fluid is therefore superconducting. It offers no resistance to the passage of an electrical current. Superconductivity was discovered in Leiden by H. K. Omnes in 1911. The pairing theory, put forward in 1957 to explain superconductivity by J. Bardeen, L. N. Cooper and J. R. Schieffer, is one of the great triumphs of modern physical theory and it earned Nobel Prizes for its proponents.

In the laboratory quantum fluids are created only at very low temperatures. Superconductors have to be cooled to within ten degrees of absolute zero and superfluid helium to even less than that. This is because the pairing energy is so weak that the thermal energies associated with temperatures higher than a few degrees break the pairing bond. Inside a neutron star the superproperties arise because of the exceedingly high density rather than because of low temperatures. The 'transition temperatures', at which superproperties take over, will prevail in all but the very youngest neutron stars. The Crab Nebula pulsar is old enough to be superfluid and superconducting.

When you stir a container of liquid, rotating masses called whirlpools or vortices form in the fluid. One unexpected consequence of neutron superfluidity inside the Crab Nebula pulsar is that the vortices cannot have an arbitrary amount of rotation. Instead the vortices are themselves restricted to certain modes of rotation, a situation that has been investigated experimentally in liquid helium. These vortices contribute to the overall angular energy of the pulsar.

An intriguing configuration of protons is probably present in a superconducting neutron star. A superconductor cannot tolerate a magnetic field in its interior. As it makes the change from an ordinary material to a superconductor it expels the ambient field. But stars, especially neutron stars, have magnetic fields inside them and it would take millions of years to expel it all. What happens is that the protons learn to live with the field by forming alternate layers of superconducting proton-pair material with no magnetic field and routine proton material able to support a field. Yet another complication in this strange quantum jungle is that the supervortices can contain some of the magnetism.

We see that the study of pulsars has led to a fruitful interaction of laboratory physics at low temperatures and astrophysics at high temperatures. It is unlikely that astrophysicists would have got so far in working out what the inside of a star is like were it not for the substantial work already carried out in the laboratory on the weird behaviour of ordinary material at very low temperatures.

6.5 Greek alphabet soup

What is it like deep inside a neutron star? Probably astrophysicists will encounter here the weird world of the elementary particles. Once the density rises above 10^{18} kg per cubic metre particles other than neutrons, protons and electrons are expected. For example, there are particles rather like neutrons and protons but somewhat more massive. The Σ (sigma) particles, for instance, can be viewed for our purposes as excited states of the neutron and proton. They are unstable at ordinary densities. Other heavy particles, or baryons as they are collectively named, may include the Λ (lambda) and Δ (delta) particles. Another class of particles, quite different in character to the baryons, is that of the mesons. The electron is a meson. Other examples which may be encountered in neutron stars are pions (π), and possibly the ρ (rho) and ω (omega) mesons. At the densities needed to create these denizens of the world of elementary particles we know very little about the properties of matter. Consequently no one can really say whether the innermost kilometre or so of the Crab Nebula pulsar will consist of these excited baryons and mesons. Nor is it possible to derive with precision the equations governing the behaviour of such a 'Greek alphabet soup' if indeed it does exist.

6.6 Models of neutron stars

The Crab Nebula pulsar in particular permits a comparison of the ideal world of the blackboard and the real world of telescopes and observatories. To do this, models of neutron stars have to be constructed on paper and in computers. The consequences of the models are then considered. Finally these conclusions from the models are assessed against the observational data.

Neutron star models lead to predictions of a range of masses and moments of inertia. Additionally details of the structure such as the stiffness of the crust and rigidity of the core are calculable. For a theorist to be able to construct a neutron star model it is necessary to have an equation of state. This is an equation which relates the pressure in the nuclear matter to its density. It is thus a relation that is derived from the properties of nuclear matter as investigated in particle accelerators (so-called atom smashers) on Earth. There is no general agreement about the precise form of the equation of state that should be used in modelling neutron stars. While this may worry some theorists it need not concern us unduly: the main effect of the different equations is to alter the maximum possible mass and certain properties of the interior. In addition to the equation of state, the equation governing the hydrostatic balance is needed. This describes the way in which the internal pressure and density combat the gravitational forces. At nuclear densities this balancing act involves another area of physics that is full of surprises: the general theory of relativity.

Calculations have shown that upper bounds to neutron star masses range downwards from 3.4 solar masses, an absolute limit irrespective of the equation of state, to around 1.7 solar masses. It is not yet possible to measure the mass of the Crab Nebula pulsar, although indirect arguments give of the order of a solar mass. However, the neutron star in the X-ray binary star Hercules X–1 has a mass of 1.3 solar masses, which is comfortably within the bounds of the models.

In the case of the Crab Nebula a further clue is obtained from the observed slowdown rate. The nebula itself radiates an energy of 1.2×10^{31} watts. At the very least the central pulsar has to replenish the electrons producing the optical and X-ray synchrotron radiation, and this demands an input of 0.8×10^{31} watts. The most plausible ultimate source of this energy is the energy liberated by the slowing down of the rotation, which can supply the entire amount if the moment of inertia of the spinning pulsar exceeds about 1.8×10^{37} kg m². (The importance of the moment of inertia is that it, multiplied by the square of the angular velocity of the pulsar, gives twice the total rotational energy.) Some models of neutron stars fail to give a moment of inertia as high as that in the Crab Nebula pulsar. Most models, however, survive this test. This is a good point at which to compare the Crab Nebula neutron star with models.

6.7 The Crab Nebula neutron star

The range of masses and moments of inertia derived in a whole series of models do not disagree with observations of the Crab Nebula pulsar. However, this unique object does furnish a critical test of theories in regard to mass and moment of inertia. This is because the mass cannot be measured directly and only a lower limit to the moment of inertia can be inferred from the argument that slow-down of the rotation keeps the nebula alight. Equations of state for nuclear matter in which the neutrons and protons do not repel each other at very close distances can be excluded because they lead to maximum moments of inertia which are smaller than those needed to explain phenomena in the Crab Nebula. Most equations of state are compatible with the energy loss seen in the Crab Nebula.

At present these models offer little more than an order of magnitude test in regard to mass and moment of inertia. But it is a different story with the glitches. These do give valuable clues on the internal structures.

Both the Crab Nebula and Vela pulsars have unexpectedly speeded up on several occasions. Two features are notable after a major glitch. Most of the frequency jump decays away exponentially after a glitch, while the relative change in the slow-down rate is always much greater than the relative change in frequency. Theorists have sought to explain these two properties by means of a two-component model of a neutron star. This model consists of an outer crust to which some parts of the interior are coupled by magnetic forces, and a superfluid interior. What happens at a glitch is the following: the outer component suddenly speeds up (we shall go into the possible reasons for this later) and an increase in pulse frequency is immediately observed. Then the interior gradually takes up the increase in angular velocity. In the Crab pulsar this takes about a week, whereas for the Vela pulsar it is a year. The relaxation times, as these periods of adjustment are termed, provide measures of the strength of the coupling between the crust and the superfluid. For the Crab Nebula pulsar the recovery time of a week is consistent with a mass of less than 0.5 solar masses, with a thin crust and a superfluid interior of protons, electrons and neutrons. The more pedestrian behaviour of the Vela pulsar suggests a more ponderous structure. The mass derived for the Vela pulsar is over 0.7 solar masses, and there is probably a solid core which lowers the proportion of neutron superfluid.

Regardless of the precise physical details, the fact that glitches heal over in weeks or even years rather than fleeting fractions of a second is the best indication astrophysicists have that the neutron matter is behaving as a superfluid. Were this not the case the outside of the star would interact far more fiercely, and abruptly, with the inside.

But why do the big glitches (or macroglitches) take place? There are several theories, not all of which are related to the structure of the star itself. The facts are these: macroglitches always lead initially to an increase in pulse frequency, they come about in only a few hours, the pulse shape is not affected, and only young pulsars are known to have made glitches.

One of the first mechanisms suggested has a terrestrial analogue. Malvin Ruderman suggested that the 1969 Vela glitch was due to a crustquake. This is a sudden release of elastic energy that has built up in the outer part of the star, and it is like an earthquake. Many processes must lead to a build up of strain in the star's crust. For example, as the spinning star slows down the centrifugal force on it diminishes. When the stress gets too much the crust cracks up in a starquake. This reduces the moment of inertia in the crust and, in order to conserve angular momentum, the crust speeds up. The same effect occurs when a pirouetting ice-skater suddenly pulls in outstretched arms: the reduced moment of inertia pushes up the angular velocity. The crust then skims over the superfluid beneath (as a skater spins over the ice) until weak coupling feeds some of the excess spin into the massive interior.

Apart from crustquakes, other ideas have received more limited support. The suggestion made in 1970, that perturbations due to a planet orbiting the neutron star affected the period, now seems untenable. There are arguments favouring unstable events in the magnetized atmosphere of a neutron star as the cause of the glitches. This appears to be a workable model for the Crab Nebula in one respect: the changing wisp structure near the star could be another manifestation of tremendous 'thunderstorms' near the pulsar. However, the model suffers from many difficulties, which it is not appropriate to discuss in detail; for example, it is hard to understand how the hypothesized thunderstorms can fail to affect the pulse shape.

Elsewhere in the sky many of the double stars that are seen to emit X-radiation do so because matter is being transferred from star to star. Could a similar accretion process of new matter onto a single star trigger a glitch? Matter cascading through a strong gravitational field onto a neutron star gains a large amount of energy. However, incoming chunks are more likely to go into orbit round the star than to knock it off course with a colossal thump. Eventually such orbital debris will drift downwards. But it is hard to see how this continuous rain could set off such a sharp glitch.

As another model we can invoke processes in the star's interior to account for glitches. Instabilities in the fluid core would upset the rotation, for example. But it is not completely certain that the pulsar has a fluid core, and it would be odd if only the young pulsars experienced the instability. A possibility that cannot be ruled out easily is that of corequakes. Here it is envisaged that quakes take place in the solid neutron core. The core is expected to possess an adequate reservoir of elastic energy which can be released at intervals in starquakes.

Finally we must consider again the restless behaviour of the Crab Nebula pulsar. We saw in the previous chapter that the pulse period displays a restless or noisy behaviour of erratic but minute variations of arrival time. Could this be due to the occurrence of microglitches? If so, then speeding up as well as slowing down is involved. It is not possible to say what causes the restless behaviour. Microquakes in the crust would probably suffice, but so would the superfluid sloshing about inside the star.

The Crab Nebula pulsar provides only a misty view into the world of neutron stars. The theoretical framework is vast, but few observational facts strengthen it. We can summarize as follows. Crustquakes give a plausible account of the glitches, but other theories cannot be totally rejected. The neutron star in the Crab Nebula is probably less than 0.5 solar masses and is ninety per cent composed of superfluid neutrons.

7 · Magnetic fields and energy flow from the pulsar

7.1 Magnetism and the cosmos

Magnetism pervades the entire Universe. Objects on all scales have magnetism associated with them, from minute fundamental particles to vast galaxies and quasars. The strength of a magnetic field may be measured in the units of either the gauss or the tesla. Although the gauss is widely used in astronomical research, we prefer the unit now used by physicists, the tesla (abbreviated T), which is one ten-thousandth of a gauss. The Earth has a magnetic field of about 10^{-5} T. Cosmic magnetic fields cover a wide range: 10^{-10} T in interstellar space, 10^{-5} T on planetary surfaces, 100–1,000 T for white dwarf stars, and up to 10^8 T in neutron stars. The fields for stars and for white dwarfs have been meaured from the effect of the strong magnetism on the spectral lines. Pulsar radio signals provide a means of finding the interstellar field. Planetary magnetism is investigated by instruments on spacecraft. For neutron stars the evidence on magnetic fields is adduced indirectly, and it is theoretical rather than observational information.

7.2 Collapsed stars and superstrong magnetism

Until the mid-1960s, theoretical work on neutron stars mainly centred on the properties of their interiors. In 1967 Franco Pacini, who had written a doctoral thesis on magnetized rotating stars, suggested that there might be a very strong magnetic field associated with neutron stars. This idea did not win immediate acceptance; indeed more distinguished astrophysicists than Pacini dismissed the theory as nonsense. A couple of years later, Tommy Gold propounded the rotating neutron star model to explain pulsars. The discovery and investigation of the Crab Nebula pulsar won over the astronomical community to rotating neutron stars. Then Pacini's very strong magnetic fields had to be taken seriously.

The basic idea behind the very strong fields is the following. When we look at ordinary stars, such as the Sun, we find magnetic fields as high as 0.01 T (100 gauss). Inside the stars the great internal pressure means that matter is ionized so there is a population of free electrons. Since the electrons carry electric currents we readily deduce that stellar interiors will be good conductors of electricity. Now try to visualize what happens when a magnetized object which is also a conductor of electricity is compressed. The magnetic force lines are actually 'frozen in' to the structure of a conducting object because of forces between the free electrons and the field. Consequently, as the object shrinks so does the field. The force lines thus become packed tighter together and the strength of the field, which can be regarded in a simple way as a guide to the density of field lines, therefore rises. If a spherical object shrinks and at the same time conserves the magnetic flux within itself, the field is amplified by a factor that depends on the inverse square of the radius. If our Sun were to shrink down to the size of a neutron star its radius would decrease by a factor 10^5 and the field would rise by a factor 10^{10}—from 0.01 T to 10^8 T. Incidentally the largest fields made in laboratories are about 10^3 T, which can be sustained through a volume of only one cubic centimetre or so.

The reasoning behind the belief that neutron stars have large magnetic fields is thus quite simple. If neutron stars are created from fairly ordinary stars in a sudden event the magnetic flux ought to be conserved in the collapse, which will inevitably lead to a superstrong field in the collapsed star. Initially, doubts as to the reality of these enormously strong fields centred on their expected lifetimes. It seemed that the fantastic magnetic flux ought to dissipate itself rapidly, rather as cheap toy magnets soon lose their magnetism. However, the continuing activity of the Crab Nebula lighthouse, as required by Gold's model, showed that the magnetic field associated with NP 0532 had not decayed. The realization that the Crab Nebula pulsar was probably superconducting furnished a convenient way of accounting for its permanent magnetism.

7.3 Magnetism at the surface of the Crab Nebula pulsar

You will not be surprised to learn that the nature of matter changes substantially in the presence of magnetic fields as large as that which must exist at the surface of the neutron star in the Crab Nebula. This is because the extraordinarily large magnetic forces acting on electrons exceed the forces governing normal atomic structure. One consequence is that atoms, instead of being spherical, become cylinders aligned along the direction of the magnetic field. Another effect is that the atoms, or rather their clouds of orbiting electrons, become considerably smaller.

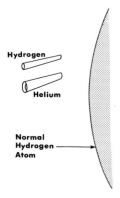

Fig. 28 Distortion of hydrogen and helium atoms in a magnetic field of a neutron star. The electron orbits become cylinders which are much smaller than the normal hydrogen atom.

Atoms in superstrong fields grasp each other extremely tightly. The distortions to the shape that the field causes lead to very strong attractive forces between atoms. It is thought that the atom-cylinders may be capable of linking up end-to-end. Thus long atomic chains, or polymers, may be found on neutron stars. For example, what may be formed are chains of iron nuclei surrounded by sheaths of electrons. These wobbly chains can attract each other and join in parallel to make magnetized iron whiskers. The matter in these threads has a density of about 10^7–10^8 kg per cubic metre—thousands of times denser than terrestrial iron. This fibrous layer on the surface of the star has a thickness of only a few metres. Below that depth the pressure is too great for them to form. So the first few metres of a neutron star consist of a thick carpet of dense iron filings.

Enormous electric fields exist close to the Crab Nebula pulsar. Whatever the detailed configuration of the magnetic field, the fact that the magnetized star is spinning thirty times a second means that there will be an electric field immediately outside the star of about 10^{14} volts per metre. This in turn implies that the total difference in voltage between the magnetic pole and equator of the star must be a million million million (10^{18}) volts. This electric field is in principle capable of accelerating charged particles, such as protons and electrons, to enormously high energies. In fact, the Crab charged particles can be boosted to the highest energies recorded for the cosmic ray particles.

Incidentally, it is instructive to compare the electric field in the vicinity of the Crab Nebula neutron star with what you can accomplish here on Earth. With a toy magnet twirling on the end of an electric drill you would be fortunate to generate more than a millivolt. Using a more expensive laboratory magnet (length 10 cm, field 1 tesla) you might generate a few volts. But nothing approaching a million million million volts can be made. For that you have to rely on the stars.

It is abundantly clear that the magnetic and electric forces completely dominate the atmosphere of a neutron star. In the Crab Nebula the ratio of electromagnetic to gravitational forces is of order a billion to one and the electromagnetic forces is the main influence. An atomspheric environment consisting mainly of electromagnetic fields and charged particles is called a magnetosphere. Our Earth has one of these above its gaseous atmosphere and extending thousands of kilometres into space. Jupiter has an extensive magnetosphere that is hundreds of times larger than the solid planet. But the most spectacular magnetospheres are found around neutron stars.

7.4 A first look at the magnetosphere

American research scientists P. Goldreich and W. Julian sketched the first theoretical account of what the magnetic envelope of a neutron star is like in 1969. They assumed that the magnetic and rotation axes were aligned (what theorists term an axisymmetric dipole), and then worked out the resulting distribution of magnetic field and charged particles. The diagram shows what the magnetic field looks like outside the star.

Electrons and protons in the magnetosphere are strongly influenced by the field. Along the field lines charges can move with ease, but motion perpendicular to such a powerful field is impossible. This means that the plasma of charged particles is locked into a magnetic cage. Although it can leak out along the bars it cannot easily trickle through them. The magnetic cage, in turn, is rigidly fixed to the neutron star. Consequently the magnetosphere of particles and fields is forced to rotate with the spinning star, and this rotation is like that of a solid body or wheel. It is easy to see that at a certain distance from the neutron star the magnetic cage and the trapped particles are being whirled around at the speed of light.

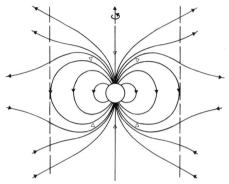

Fig. 29 The magnetic field outside a neutron star. Solid arrows show the direction of the magnetic field. Open arrows indicate the direction of flow for electric currents in the nagnetosphere. The dotted line indicates the 'speed of light' cylinder beyond which the field lines must sheer and break in reality.

For the Crab Nebula pulsar this distance—the speed of light cylinder—is 1,500 km from the centre of the star. Beyond this cylinder the magnetosphere cannot spin round in unison with the central star, but there is no agreement among astronomers about physical conditions in this vicinity. As the particles approach the velocity of light their mass increases greatly, in accordance with the theory of relativity. This increase in the inertia will cause the field lines to trail behind the rotation. From the diagram you can see that some field lines, those near the equator, close back in on themselves. These lines enclose a volume within which electrons are trapped. Outside this region there is an excess of protons. Near the two magnetic poles particles can stream away from the star along the field lines. This region through which particles can escape is known as the wind-zone of the neutron star. Open field lines allow particles to drift through the speed of light cylinder, although they do not, of course, exceed the speed of light as they do so. This leakage of particles leads to a transfer of angular momentum from the spinning star to charged particles further out. This leakage of energy contributes to the slowing-down of the spinning star. At the speed of light cylinder the magnetic field is still about 100 tesla (i.e. one megagauss) for the Crab Nebula pulsar.

7.5 Complications in the magnetosphere

The simple model of Goldreich and Julian is for a magnetic axis that is parallel to the rotation axis. What will happen if the internal magnet is tilted at an angle to the rotation axis, as is the case for the planets Earth and Jupiter? Fortunately much of the physics is unaltered. As in the simple model there are large electric fields that strip charged particles away from the surface. An important difference is that the spinning magnet radiates magnetic dipole radiation. At the light cylinder the magnetic field fluctuates thirty times per second as the star spins, and electromagnetic effects exert a torque on the star that is as important as the effect of the outflowing matter.

7.6 Energy flow from pulsar to nebula

In Chapter 5 the point was made that the observable parameter known as the braking index can be used to discriminate theories. This number is a measure of the strength of the braking being applied to arrest the rotation. For a pure dipole field (like a simple bar magnet) rotating in a vacuum the braking index is 3.0. For the Crab Nebula pulsar the index is not quite as high as this (2.6). Clearly there are complications introduced by the presence of charged particles in the magnetosphere which modify the field from a simple dipole. Nevertheless the agreement is close enough to show that the simple models for the Crab Nebula pulsar as sketched here must be along the right lines.

From the oblique rotator model it is possible to make a rough estimate of the number of electrons that cross the speed of light cylinder. Certain further assumptions about how the electrons and protons are distributed in the magnetosphere have to be made. The expected particle loss-rate then comes out at 10^{36}–10^{40} particles per second. This quantity agrees neatly with the rate at which the Crab Nebula itself is continuously losing energetic electrons through synchrotron radiation. Furthermore the energy of the individual electrons produced by the acceleration mechanism is calculable. There are no serious difficulties in producing electrons in the energy range 10^8–10^{14} volts (eV) by magnetic dipole radiation. Again this agrees well with the energy distribution that is needed to maintain the nebular synchrotron radiation. Finally, the total energy flux across the speed of light cylinder can be made to agree with the energy flux from the nebula without difficulty. Hence we see that several features of the energy flow from the spinning neutron star concur with the input needed to maintain the nebula. This balancing of the energy budget strengthens the belief that the basic models of particle production are correct.

Near the magnetic poles the outflowing charged particles are rapidly accelerated to extremely relativistic energies. Then an important process takes place which greatly increases the density of charged particles. The relativistic electrons, being charged particles, are constrained to move along the curved magnetic field lines. They emit a type of radiation known as 'curvature radiation'. This name arises because the basic laws of physics show that an electric charge must emit radiation when it is accelerated; motion along a curved path is an accelerated motion. The curvature radiation is emitted as gamma-rays. These gamma-ray photons each have more energy than two electrons; as a consequence of their high energy and the strong magnetic field they produce electron–positron pairs. You can imagine that the positron is basically a positively charged electron, so the created pairs are electrically neutral overall. The strong magnetic field separates them before they can re-unite to make a photon.

Now a crucially important 'cascade' effect comes into play. The new electron and positron are accelerated by the field. In turn as they flow down the field lines they emit photons of curvature radiation which spontaneously spawn electron–positron pairs. And so the chain reaction goes on, providing a copious source of high-energy charged particles. Ultimately the energy to make these particles

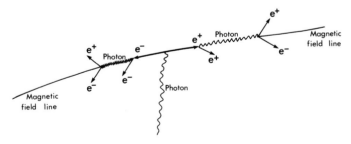

Fig. 30 Production of high-energy particles by cascades. The photons emitted decay into pairs of charged particles which then scatter to produce further high-energy photons in collisions. These in turn decay into particle pairs and in this way a large population of charged particles is produced.

must come from the slow-down of the star. We are now beginning to get an insight into the workings of the particle accelerator inside the Crab Nebula, which feeds the rotational energy of the neutron star into new clouds of charged particles.

Once the cascading particles move a certain distance from the neutron star the field is no longer up to strength for making curvature radiation at high frequencies. More important still, as a neutron star ages the field dwindles. In the curvature radiation picture the emission of optical and X-ray flashes cuts out sharply as the field dies. Also the energies and number of particles decrease rapidly as the pulsar spins down, and this also removes any possibility of optical pulses. So the curvature radiation model enables us to understand why the youngest pulsars are the jazziest.

Where does the pulsar magnetosphere end and the Crab Nebula begin? Once the energy has crossed the speed of light cylinder it is cut off from the spinning neutron star, and diffuses out with spherical symmetry. The magnetic field falls inversely with increasing distance, reaching a value of 10^{-8} tesla (10^{-4} gauss) about one light year from the star. This agrees with the strength of the field in the amorphous part of the nebula. Some theorists consider that the neutron star probably sustains the magnetic field throughout the nebula.

7.7 The nebula shines on

A general consensus emerges from observations of the Crab Nebula and pulsar NP 0532. The nebular magnetic field is 10^{-8} tesla and the total stored energy in relativistic particles is 10^{42} joules.

Continuous injection of new particles into the nebula must take place in order to sustain the spectrum of synchrotron radiation. Where can these particles come from? The simplest explanation is that the central pulsar keeps the nebula going. Over the lifetime of the object it has amplified the magnetic field and fed in fresh doses of fast electrons. Now we want to take a closer look at the link between the energetic pulsar and the glowing nebula.

The Australian scientist J. H. Piddington suggested in 1957 that the magnetic field in the nebula might have its seat in a massive rotating body. Subsequently (1965) the Soviet theorist N. S. Kardashev indicated how the magnetic field around a rotating star could be wound up and thus intensified. Pacini suggested in 1968 that a neutron star could furnish the necessary supply of relativistic particle energy. All this research happened before NP 0532 was found. Since then Martin Rees and James Gunn have given (1974) one picture of how the magnetic field and particle energy may be sustained in the nebula.

Rees and Gunn adopted the inclined rotator model for the direction of the magnetic field. They pointed out that three distinct types of energy will flow through the speed of light cylinder. These three types are: (1) a wind of fast particles, as mentioned in section 7.6; (2) a magnetic field, shaped like a ring doughnut, that is carried with the wind; and (3) low-frequency (30 cycles per second) electromagnetic waves. The effect of the tremendous stellar wind and associated magnetic field in the vicinity of the pulsar is to sweep out the charged particles from a spherical cavity of radius about one-third of a light year. At this distance the wind is seriously dissipated. Its pressure matches the internal pressure of the nebula, and so no further removal of charged particles can take place. At the boundary of this hole in the interior of the nebula the low-frequency (30 Hz) waves are absorbed and their energy is transferred to relativistic electrons. Possibly, though not certainly, the boundary of the cavity is related to the innermost series of the moving wisps studied by Lampland and Scargle. The amorphous continuum of the nebula almost vanishes close to the pulsar, showing that a genuine hole does exist at the centre.

At the cavity boundary a highly disturbed region develops because the relativistic wind, moving at over half the speed of light, crashes into the almost stationary bulk of the nebula. Beyond this shock zone the velocity of the wind decreases smoothly from half the speed of light (i.e. 1.5×10^8 metres per second) to the expansion velocity of the nebula (observed to be about 1.5×10^6 metres per second). Rees and Gunn envisaged that the doughnut-ring field dragged along with the wind would be amplified as the wind slows. Remember that magnetism connected with a good conductor is essentially frozen in and must move with the conductor. As the wind slows the conduction medium is being effectively compressed, and the consequential crowding together of the electrons intensifies the associated field. So although the outflowing magnetism may not be strong even near the pulsar, it gets amplified in the nebula as the stellar wind slackens. In this model some of the electron wind and magnetic field overshoots the optical boundary of the nebula. This explains why the nebula looks larger at lower frequencies. The close relationship of nebular magnetism to electron pressure in this model also accounts for the irregular outline of the nebula because the magnetic stresses must induce unstable situations in the conducting plasma. The continuous radiation emitted by the nebula at all frequencies from radio to gamma-ray is due to the synchrotron process.

One of the great achievements in the study of the Crab Nebula has been the linking of the pulsar energy to the luminosity of the nebula. The explanation of the energy budget was a crucial advance in understanding the physics of the Crab Nebula. Also the agreement of the models with the observations is unusually good. Once again we are conscious of the Nebula's unique qualities. It is fired by a star which is itself invisible. A star that derives its energy not from nuclear burning but from the braking of a simple flywheel. The vast filaments of glowing gas, beckoning spectroscopists, are lit by a star only ten kilometres in diameter. The Crab Nebula is indeed a remarkable physical object.

8 · How does the pulsar pulse?

8.1 The problem

We believe that the spinning of a neutron star provides the clock mechanism for a pulsar, but this simple concept offers no insight into the emission mechanism itself. What astronomers want to know about pulsar emission may be summarized in the following questions: where is the emission region located relative to the neutron star; how can the Crab Nebula pulsar perform over so many decades of frequency; how is the spectrum produced; and what gives the pulses their distinctive shapes? Answering these questions is very difficult, and it is not easy to explain in an elementary way the partial solutions derived so far. This is an area of astronomy that intimately links electrodynamics, the special theory of relativity, plasma physics, and magnetism. The links mainly consist of mathematical relations. Working out how a pulsar functions is rather like trying to deduce how a TV station works merely by looking at the programmes from a distance of 1,000 parsec.

The astronomical journals are littered with articles that 'explain' features of pulsar emission, using a variety of radiation mechanisms. It is hard to find a simple route through this forest of ideas. We stick to the known and observed facts and show in a general manner how some of them may be accounted for. The models and theories described here reflect somewhat the author's own prejudices, although they do also have a reasonably wide following among other astronomers.

In many respects the Crab Nebula pulsar sets the severest constraints upon theorists. For example, it fires out photons with the energies of radio waves as well as gamma-rays. It might be easy to make radio pulsars; but the Crab pulsar sets a much tougher problem. However, the other pulsars have contributed much useful information, particularly upon the range actually encountered in the magnitudes of various quantities such as periods and slow-down rates. In constructing a model of pulsar emission we draw on the abundant data available for all the pulsars, not just the youngest and fastest example.

8.2 The data

Pulsar emission theory is still in the embryonic stage at which the data seemingly run ahead of the ideas. The main points to keep in mind when embarking upon flights of fancy are these:

(1) For most pulsars the pulse width takes up about three per cent of the total cycle. This means that the lighthouse beam sweeping the skies is about 10° broad on the average. This width can most conveniently be regarded as a measure of the size of the emitting region. For the Crab pulsar the width is more or less the same at all observed frequencies.

(2) Considerable stability is shown by the average pulse profiles, which are often the same year after year. Hence the emitting region also must have an essentially stable structure.

(3) The polarization within a pulse is usually linear and it sweeps smoothly through an angle of perhaps 100° across the pulse. This behaviour is probably connected with the direction of the average magnetic field near to or within the emitting region.

(4) When it comes to looking at a pulsar's individual pulses there are enormous variations. It is only the average behaviour of many pulses that is steadfast: each pulse possesses its own personality within this norm.

(5) Some pulsars, including the Crab, exhibit an interpulse. In general this does not occur halfway between successive main pulses.

(6) The spectrum of the pulsed radiation follows the power-law behaviour that is typical of radiation by the synchrotron mechanism. In the Crab Nebula the optical and radio pulsed spectra join smoothly but the infrared spectrum declines at lower frequencies. The radio spectrum does not obviously tie in with the infrared, optical or high energy spectra, despite the close similarity of the average pulse profiles at these different energies.

(7) The periods of over 160 pulsars span a factor of a hundred, from 33 milliseconds (Crab pulsar) to 3.75 seconds (0525 + 21). The slow-down rates spread through four decades from 36 nanoseconds per

day (Crab pulsar) to about 0.001 nanoseconds per day (1813 − 26). An important fact is that a graph of the period against slow-down rate for all pulsars does not show any obvious connection between the two quantities. This lack of correlation means that the pulsars cannot be arranged into a neat evolutionary sequence based on the observed periods. Despite this we can still argue generally perhaps that the fastest pulsars are young in some sense and the ponderous pulsars are old. Even though the rotating neutron star hypothesis implies that rotation is linked to the emission of energy, there is no strong correlation of periods or slow-down rates with luminosities. The only clear-cut observation is that there are no very slow pulsars. By the time the star has slowed to one rotation per second its radio emission has evidently reached the point where it will soon be turned off.

(8) In the graph of period (p) against slow-down rate ($\dot{p}$) there are no pulsars beyond a line at a constant value of $\dot{p}/p^5$. This fact is most intriguing. For a spinning star $p\dot{p}$ is proportional to the magnetic field at the surface, and $\dot{p}/p^5$ is proportional to the square of the field at the speed of light cylinder—the surface where a body forced to rotate in unison with the star would be moving at the speed of light. This suggests that pulsar emission is related to the speed of light cylinder, and that the signals die when the field (indicated by $\dot{p}/p^5$) falls below a certain critical value.

(9) Only two pulsars, the Crab and the pulsar in Vela, are seen optically. Both are young, fast pulsars. The Vela pulsar is so faint that it seems implausible that flickering light will be detected from any other slower pulsars.

The above general remarks must guide any investigation of emission mechanisms. Unless a model is consistent with these points its ability to explain more esoteric observational minutiae must be considered only as an academic exercise.

8.3 The seat of activity

The lighthouse model of Gold is easy to grasp. Every rotation of the ultra-dense star sweeps a lighthouse beam over you. A disadvantage of its intrinsic simplicity is that it may lead us to feel that the radiation is beamed at us by a natural light (star spot? flare? plasma instability? magnetic pole?) on the surface of the neutron star. We must be more scientific than this. How can we find out where the pulse is generated?

Most of the useful information that we can bring to bear arises from points (1), (2), and (3) in the preceding section.

We can fairly readily rule out unpredictable irregularities such as star spots. If these blemishes existed they would presumably vary markedly in size from neutron star to neutron star, and then so would the pulse width. The constancy of the average pulse width among the pulsars shows that a more general property of the star and its magnetic field must be implicated. We seek a small and closely-defined region because only then can our model satisfy (1), (2) and (3). One obvious candidate is the polar cap, where the magnetic field lines converge to produce a very intense magnetic field. Other less obvious regions will be considered below.

Some properties of pulsars are so simple that they scarcely make an impact on us. Take the average pulse profiles which we are discussing. Generally these are very symmetrical about the peak intensity. From this we conclude that the emission region also has an underlying geometrical simplicity.

The behaviour of the linear polarization, swinging through up to 100° during a pulse, leads to two possibilities. Either many emitting regions with slightly differing polarizations are viewed in rapid succession as the pulse sweeps past the observer, or one source with one fixed polarization is viewed from different aspects as the star rotates. In practice both phenomena may well be at work. The single-source model has been applied to the Crab Nebula pulsar. This suggested that the rotation axis and magnetic axis of the neutron star almost coincide; however, it is not certain that the model is applicable.

Clearly, if we are to proceed scientifically, we must also consider if it is possible for the emitter to be well away from the surface of the star. Whereabouts is this likely to be? Proceeding outwards the first physically important region that is encountered is the speed of light cylinder. If remarkable events are to happen then this extreme boundary is a good place to start the calculations.

Whereabouts on the speed of light cylinder might we expect to find a concentration of high speed particles and magnetic field? A diagram is helpful to this discussion for it shows that there are two regions where the field lines run parallel to the speed of light cylinder. Particles here might be trapped just inside the cylinder by the magnetic field. Another possibility is argued by Graham Smith: charged particles want to flow along the field lines, but they cannot exceed the velocity of light; therefore, when a particle sliding outwards along the field encounters the speed of light cylinder it must acquire enough energy to break free of the field lines. Calculations indicate that the region on the light cylinder where this occurs extends for about 10°, which neatly matches the observed pulse width. It is hard to be more specific than this because no one has succeeded yet in modelling the geometry of the magnetic field close to the light cylinder. One crucial feature of models that positions the source far out in the magnetosphere is that the source is then orbiting the neutron star at close to the velocity of light. This opens up the possibility of effects in special relativity assisting with some of the problems.

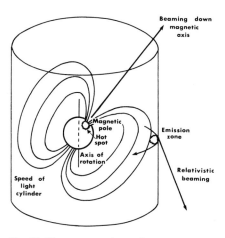

Fig. 31 The magnetosphere of a neutron star showing the region near to the speed of light circle where sources of radiation may be located, as well as the hot spot at the magnetic pole. In reality only one of these two postulated emission regions can be responsible for the observed pulses.

For the Crab Nebula pulsar a remarkably simple energy argument positions the seat of activity well away from the neutron star. The argument runs as follows. Suppose that the active region is on the surface, whose total area is known. Then, even if the entire surface is involved the minimum temperature of the emitting region needs to be 10^{15} K in order to produce the observed brightness. We need to explore the consequences of such a high temperature. We can calculate the corresponding thermal energy of charged particles, which comes out at a value approaching one million times the rest energy. At this high temperature the velocity of the electrons is very close to the speed of light. Knowing the velocity from the implied amount of thermal energy we can next calculate the orbit of one of the electrons in the pulsar magnetic field. (Remember that electrons gyrating round field lines are the cause of the synchrotron radiation.) In fact the electrons would orbit on paths with a radius of about 10^5 km, which is vastly greater than the radius of the neutron star. Therefore they travel essentially in straight lines and would lose only one-billionth of their thermal energy before leaving the star. This is much too low. Other arguments show that a good fraction (much more than one-billionth) of the rotational energy loss from the neutron star is fed into the pulsed radiation. A typical pulsar is losing 10^{26} watts of rotational energy but radiates only 10^{20} watts as radio waves. We see, therefore, that our assumption of pulse energy generation close to the surface leads to inconsistencies with theory and observation. We must discard the hypothesis in this form. It can be rescued by postulating that the electrons are accelerated in coherent bunches that would act essentially as a single particle with the combined mass and charge of its constituent electrons. Then the emission mechanism becomes far more effective. The Crab Nebula pulsar proves that the pulses are not beamed from the surface itself, unless the pulse is made by coherent radiation.

Although there is agreement that starspots probably cannot make the pulses, opinion is divided on the location of the activity in the magnetosphere. We have already mentioned the speed of light cylinder, but what about the polar caps? No thorough model exists for the magnetic field of a spinning neutron star with a magnetic dipole inclined at an angle to the rotation axis. Nevertheless many papers have been published, based on a simple polar cap theory. The idea here is that interesting physics can happen when the field lines are funnelled together at the magnetic poles. One prediction of polar cap models is that the width of the pulse should depend on period. No such correlation is observed in practice, and this therefore goes against the model.

According to the polar cap model, emission that we see as pulses is generated as electrons stream from the stellar surface along the widening funnel of magnetic field. The radiation mechanism is thus that of curvature radiation (electrons streaming along curved field lines) and not synchroton radiation (electrons orbiting around field lines). Observations do not give strong support to this mechanism.

By default we are left with a location at the light cylinder. Other positions do not work so convincingly. However, an open mind must be kept on these problems. If good models are developed for an inclined dipole it might well be the case that irregularities in the field structure will lead to a concentration of electrons and the existence of suitable sources away from the light cylinder. Some preliminary models have suggested, however, that such regions of interest are only found out at the light cylinder even for the inclined dipole.

For the rest of the discussion we assume that the active region is in fact locked into the magnetic field at the distance where matter is orbiting on the field lines at close to the speed of light.

8.4 Relativity makes a searchlight

Many people who have not made a study of the special theory of relativity are, nevertheless, aware that this theory gives predictions about the behaviour of matter moving at high speeds relative to the observer, and that these predictions are not intuitively obvious. In fact, it is precisely because the predictions of relativity theory are not intuitively obvious that the theory is baffling to many without scientific training. One of the predictions of the theory is that the mass of an object as measured by an observer will increase greatly as the velocity of the object relative to the observer approaches the speed of light. Another deduction is that material objects cannot be accelerated indefinitely. They cannot be made to pass through the speed of light barrier because they would then have an infinite mass, and therefore infinite kinetic energy, to get through. You can explore these facts and many others in the good popular books that have been written on the physics of space, time, and relativity. In this section we want to take a brief look at the implications of relativity for our model of pulsar emission.

Recall that one requirement of viable pulsar models is a method for confining the radiation so that something akin to a lighthouse beam is produced. For the Crab Nebula pulsar we know that this ray of photons has more or less the same angular width at all frequencies from radio waves to hard gamma-rays. How are the beams of radiation made on Earth? In the case of radio and television transmissions large antenna masts are used, equipped with arrays of dipoles or reflecting dishes. The problem of making a narrow beam of highly polarized radiation is well-known in radio engineering. All the man-made solutions have a common property: at longer wavelengths the beam becomes broader because diffraction takes place at the transmitting antenna. In this context, diffraction may be viewed as a broadening in the directivity of the beam; it is caused by the wave nature of electromagnetic radiation. Radio engineers cannot make a system of antennae that would produce a beam whose angular size did not increase with wavelength. When the wavelength is compara-ble to the dimensions of the transmitting dipole the beam is about 60° wide, for example.

We see that we must find a way of concentrating light into a beam that is essentially different from the methods used by radio engineers.

Relativity theory offers a means of doing this. On Earth the observer (your television, say) never moves at an appreciable velocity relative to the source of radiation (say a TV transmission mast) and hence the theory of relativity is superfluous to a discussion of terrestrial radio engineering. But in the universe we certainly do have the possibility of sources travelling very fast relative to our telescopes. In pulsars this is a probability. There must be material at the speed of light cylinder: if this matter is radiating then special relativity must be brought into the analysis.

Several researchers, including Graham Smith of the Royal Greenwich Observatory, have shown how relativistic effects will produce pencil beams of radiation. Consider a source of radiation that is moving at around ninety per cent of the speed of light. Let's not bother why it radiates because the analysis is true no matter what mechanism is stimulating the emission.

An extraordinary fact is that relativistic motion compresses a beam of radiation so that it looks narrower to an observer than it does to the source. For a speed of ninety per cent the velocity of light this narrowing is by a factor of just over 2; at ninety-five per cent of the velocity of light the factor exceeds 3. In addition there is another effect at work which is associated with the nature of time. When the emitting region is heading towards the observer a time compression takes place and this reduces the apparent width by rather more than the factors mentioned above. Putting both effects together, as we should, shows that the compression is by a factor of 10 at ninety per cent the speed of light and by a factor of 35 at ninety-five per cent.

A full analysis shows that this relativistic beaming is quite capable of accounting for the observed beam widths. The observed widths of sub-pulses are in the range 1° to 3°, which requires a speed of ninety-five per cent the light velocity. A further interesting property of this mechanism is that the beam is fan-shaped rather than pencil-shaped. The pulse is seen as the fan flicks past the Earth. A typical pulsar fan beam sweeps through one-third of the sky. Reversing the argument we can predict that we have already seen about one-third of the pulsars that would be seen if the rotation axes of all of them were inclined to ensure that the beams intersected the Earth. Clearly any deductions about the galactic space density of pulsars will depend strongly on the beam-shape because we have a much better chance of detecting a fan beam than a pencil beam.

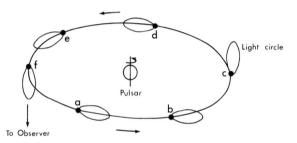

Fig. 32 The beaming of radiation from a source moving close to the speed of light. The observer far from the orbiting source sees a flash of light once during each orbit. The high velocity has the natural effect of concentrating all the radiation into a narrow beam directed along the instantaneous direction of motion.

8.5 Polarization

The location of the emitting region at the light cylinder is consistent with the observed behaviour of the polarization inside a pulse. When a pulse or sub-pulse is dissected electronically it is found that the principal plane of polarization rotates smoothly during the duration of the pulse. An analysis has shown that the smooth variation of the polarization indicates an underlying simplicity and uniformity in the magnetic field in the neighbourhood of the emitter. In fact these polarization data can even be interpreted to tell us where the emitters are. According to one piece of such detective work they are to be found where the open field lines cross the light cylinder.

8.6 Spectrum of the pulses

Nobody seriously doubts that the pulses from the Crab Nebula in the infrared, optical, and high energy wavebands are due to synchrotron radiation because the spectrum throughout that range is entirely consistent with this picture. Even the drop in intensity in the infrared can be accommodated by the mechanism of self-absorption: at the infrared frequency the electrons actually take back the energy emitted by higher-speed electrons! For the Crab pulsar the synchroton model yields this value for the magnetic field strength: 10 tesla (10^5 gauss). This is so far below the value at the star's surface (where it would be 100 million times stronger) that we have further confirmation of a location remote from the star. An electron density of one billion (10^{12}) per cubic metre is required, with electron energies exceeding 300 million electron volts.

The pulsed radio radiation from the Crab pulsar is probably not due to the synchrotron mechanism. Rather it seems likely that bunches of electrons cooperate to gyrate round the field lines as self-contained units. This will lead to the coherent emission of yet another type of electromagnetic radiation: cyclotron radiation. This coherence is demanded by the high intensity of the radio pulses. On Earth such coherent radio emission can be made in a maser. However theorists have not succeeded in putting together a model of the Crab pulsar that is based on maser action. Coherent cyclotron radiation therefore is the most plausible explanation.

8.7 Interpulses

Interpulse components, seen in the Crab Nebula pulsar, are a feature of several objects. How can these be fitted into the theory of the magnetic environment of a neutron star? If we adopt an inclined magnetic dipole as our model, then two pulses should be observed per rotation. In general these will not be of equal strength because the opposing fan beams will not be seen from the same aspect. But why do many pulsars not have an interpulse?

An answer to this problem may be sought in the Crab Nebula pulsar. At optical frequencies the situation is symmetrical, but at radio frequencies the precursor pulse occurs. Maybe this tells us that coherent cyclotron radiation (radio mechanism) is more aware of small differences in the magnetic field structure than synchrotron radiation (optical and high-energy mechanisms).

9 · Outburst and aftermath

9.1 Types of supernova characteristics

What sequence of events created the Crab Nebula in A.D. 1054? What kind of explosion was it? Why did it happen? How common are these stellar cataclysms? We need to answer these intriguing questions in order to place the Crab Nebula in its correct astronomical context. One way of doing this is to look in a general way at the information derived from the study of other supernovae. As we now know those 'nova' stars that fascinated Kepler and Tycho, the greatest astronomers of their day, must have been supernovae in fact. By an ironic twist of nature no further supernovae have been noticed in the Milky Way since the invention of the telescope.

It was Hartwig in 1885 who studied a new star (S Andromedae) that suddenly appeared in the Andromeda Nebula. This object is a great galaxy similar to the Milky Way. Hartwig's first telescopic work was observation of a supernova, although its significance was not appreciated in the nineteenth century. When S Andromedae lit up nobody knew the correct distance to its parent galaxy, so its intrinsic luminosity could only be guessed. The correct calibration of the distance scale for the galaxies eventually showed that S Andromedae had flared up in luminosity far more than an ordinary nova. It was so intrinsically brilliant that astronomers realized outbursts ought to be visible in many of the nearby galaxies.

Fritz Zwicky of the Mount Wilson Observatory, California, had the commendable foresight to commence an organized patrol of new exploding stars by regularly photographing local galaxies. This search still continues today. About a dozen supernova outbursts in other galaxies are logged every year. A few are found early in their outburst and hence are useful for follow-up studies. The main information sought for a new supernova is the light curve. This is a graph which

(a) 1937 Aug. 23. Exposure 20^m. Maximum brightness.
(b) 1938 Nov. 24. Exposure 45^m. Faint.
(c) 1942 Jan. 19. Exposure 85^m. Too faint to observe.

14. A supernova flares in spiral galaxy IC 4182.

displays the evolution of the brightness with time. It is not possible to detect radio emission or X-rays from new supernovae.

From the light curves of many supernovae Rudolph Minkowski discriminated two principal types of supernova. Subsequently further classification has been attempted, but this specialization need not concern us here. The more homogeneous of the two principal classes is Type I, which is distinguished by a light curve that has an initially-rapid rise to the maximum followed by a rapid decline and then finally a slow decline lasting many months. The maximum absolute brightness is about -19 magnitudes for a Type I supernova. S Andromedae for example, reached -18.6 mag, at which point it contributed about ten per cent of the total light from the galaxy M31. The supernova of Type II are more of a mixed bag; their light curves show a slower rise to maximum and a more leisurely decline. Some astronomers consider that spectra furnish a more reliable distinction between the types of supernovae than the light curves. Here the difference appears to be quite sharp: Type I spectra do not clearly show the Balmer series of hydrogen in emission whereas Type II supernovae do. Zwicky considered that there were three more types of supernovae, but this opinion is not generally accepted today.

There is a striking astrophysical difference between the two classes of supernovae. Unfortunately there is also a confusing clash of nomenclature! Type I supernovae are associated with the old stellar component of galaxies (called population II), whereas Type II supernovae occur in the young stellar component (called population I). The old population does not contain very massive stars: these stars evolve rapidly and they exploded long ago. Therefore the Type I supernova, which is seated in an ancient object, is an explosion of stars with mass below 1.2 solar masses. On the other hand Type II

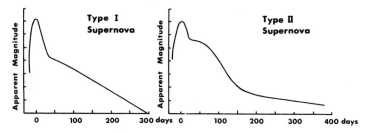

Fig. 33 The average light curves of Type I and Type II supernova explosions.

explosions are seen among the young stars and their spectra indicate significant expulsion of gas. Values of up to 10 solar masses for the precursors of Type II supernovae do not seem unlikely.

Was the event of A.D. 1054 a Type I or Type II explosion? The oriental records indicate the slow and steady decline typical of Type II outbursts. After all it took the best part of two years to fall below naked-eye visibility; the long period of daylight detectability is hard to reconcile with the sharp rise and fall of a Type I supernova. Therefore we conclude that the Crab Nebula is the relic of a Type II supernova. Of course, it is always possible that the Crab Nebula explosion was in some way very rare or even unique.

This latter possibility has received attention. In 1971 Minkowski claimed that the Chinese observations do not accord with either of the main types of supernova. But he partly based his argument on the false notion that Type II supernovae are characterized by rapid decline. The Type II supernova seen in the galaxy NGC 1058 in 1969 matched the slow decline of the light curve of the Crab supernova very closely. This example shows that the Crab event does resemble Type II behaviour.

The present shape of the Crab Nebula is a real puzzle because it does not resemble any other remnant. While most supernova remnants are delicate shells of cosmic tracery, the Crab Nebula is still a compact cloud whose interior is filled with high-energy electrons. Will it too eventually be a shell? We do not know.

A smattering of evidence suggests that the Crab Nebula is a member of a small and discrete group of rare remnants. Radio astronomers have found two other ghosts of exploded stars that look superficially like the Crab Nebula. These are the radio sources 3C 58 and G 21.5–0.9 3C 58 is the radio source left by the supernova of A.D. 1158. We must keep open the possibility that the Crab Nebula is neither a Type I nor Type II object. But for definiteness we will assume that the explosion at least was Type II, even if the after effects were somewhat abnormal.

9.2 Type II supernova explosions

We cannot travel back in time with modern instruments and observe the A.D. 1054 explosion. Instead we have to monitor contemporary explosions that look similar to it. Unfortunately these are so rare that one can only study them by looking at them when they occur in other remote galaxies. They are rather faint so that large telescopes are needed.

Statistical arguments based on the observed frequency of supernovae, their location within galaxies, and the type of galaxy in which they are found, can be used to tie down the mass range for the stars that give rise to Type II supernovae. Further refinement of the mass range comes from computer modelling of the final evolution of massive stars and of the supernova explosion itself. These two lines of investigation show that the ancestors of Type II supernova are mainly between 3 and 8.5 solar masses.

The broadening and Doppler shifting of lines in the spectra of Type II supernovae show that the expansion velocity of the supernova envelope is initially in the range 10,000 km/sec before settling to an average of 5,000 km/sec. The underlying continuous radiation is typical of that from a black-body (a perfect emitter of radiation) at a temperature of 5,000–10,000 K. The peak luminosity of a Type II supernova is typically 10^{35} watts, which rivals the output of a small galaxy. The total radiated power is around 10^{41} joules. Type I supernova are a little more energetic than this. The energy value of 10^{41} joules is comparable to the total energy radiated by our Sun in 15 million years. To this optical energy we must add the kinetic energy of the material blasted into space. In order to do this an estimate of mass of the debris is needed, because the kinetic energy is 10^{42} joules per solar mass of ejecta. The expanding remnant of the Crab Nebula lies between 1 and 10 solar masses. This range gives a lower limit of 10^{42} joules, so the kinetic energy released is equivalent to the solar luminosity over 150 million years.

Where does this immense supply of energy come from? Only two sources of energy can supply the tremendous amounts needed sufficiently quickly: nuclear reactions and gravitational collapse. It is likely that both occur in practice.

9.3 Before the Crab Nebula

Observations of extragalactic supernovae as well as theoretical studies have given a picture of the final run-up to the explosive death of a massive star. This is a field in which the theorist still reigns supreme because there is no example of a star being photographed or otherwise investigated before it became a supernova. We do, however, know that Type II supernovae predominate among the youthful population and it is primarily this fact that gives the mass range 3 to 8.5 solar masses for the progenitors. The evolution of such stars can be explored theoretically in considerable detail.

At some time in the past the supernovae were normal stars, burning hydrogen to produce helium with a consequential release of energy in a thermonuclear reaction. The details of this process, which sustains stars like the Sun, are well understood. The heat inside a star's core is sufficiently high and the density sufficiently great to enable hydrogen nuclei (protons) to fuse and to form helium nuclei (alpha particles). In this nuclear reaction considerable energy is released. A helium nucleus has 0.5 per cent less mass than the four hydrogen nuclei that made it: some matter actually disappears as a result of the nuclear reaction. This mass loss appears as radiant energy, and ultimately as starlight. The life of a star is a surprisingly delicate balance between the inward pull of gravity and the dispersing effects of the internal heating. The nuclear reactor inside a star provides the heat that is necessary to keep up the internal pressure and so prevent the star's gravitational collapse. If for some reason this reactor is turned off the star begins to cool, which reduces the gas pressure. This in turn leads to shrinkage of the stellar interior because the pressure support has been reduced. A catastrophe is now about to occur as the star plunges into its own gravitational field.

The collapse can only be halted by restarting the nuclear reactor or by the appearance of a new source of pressure. As a star runs out of nuclear fuel its inside will inevitably collapse. But why, as a sequel to implosion, do we see an explosion and the ejection of the outer parts of the star? To anwer this we must look more closely at the pathology of a massive star, depleted of its energy.

When a massive star has used up the central supplies of hydrogen it consists basically of a helium core (the ashes of the reactor) surrounded by a thick shell of hydrogen (unused fuel) in which the density and temperature are too low for the ignition of nuclear burning. The core now starts to contract. This process releases gravitational energy that appears ultimately as heat. Heat flowing out of the core warms the hydrogen next to the core and soon sparks hydrogen burning in a thin shell immediately next to the core. After this has happened helium made in the shell is added to the core, which continues to shrink even though its mass is increasing. Meanwhile the envelope of the star expands. An observer paradoxically sees the size of the star increase as the core contracts!

In the massive stars the helium contracting core can increase in density enough to trigger a new source of energy: the fusion of three helium nuclei to make a nucleus of carbon. Helium burning thus takes over the main task of propping up the star, and a carbon core is soon produced. When the helium burning is exhausted the star has a double-shell structure: inert and contracting carbon core; helium-burning shell; and hydrogen-burning shell. Nuclei are passed through the shell reactors until they end up on the carbon core. What happens next? There may be a phase in which convection in the outer part of the star stirs up the hydrogen shell. At this stage the star may be observable as a red giant. But leaving this aside, the next process is essentially that the growing ball of carbon can eventually burn to synthesize oxygen, neon, sodium and magnesium. More concentric nuclear furnaces, hotter and hotter as we move toward the heart of the star, are set up. The neon and oxygen can in turn react to make elements in the range magnesium to sulphur. Finally, if the core heats up enough during its contraction phases, silicon burning commences, with the consequential production of the elements near to iron in the periodic table of the elements.

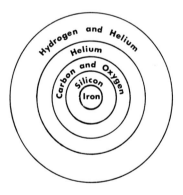

Fig. 34 Shell structure of a massive star nearing the end of its evolution. Massive stars may be an important cosmic source of the heavier elements.

If these ideas are correct then the carbon present in the ink on this page was cooked inside a stellar oven at a temperature of 200 million degrees and the iron atoms in the printing press were synthesized inside a cosmic furnace operated at three billion degrees! Even the oxygen that we breathe was probably manufactured inside pre-supernova stars at several hundred million degrees. Almost every atom of your body was made inside an exploding star aeons ago.

Ultimately the successive cycles of core contraction, envelope expansion and shell-burning result in a core that is sufficiently dense that it becomes degenerate. This core is thus supported by the pressure of degenerate electrons and thus bears resemblance to a white dwarf star.

Meanwhile material rains down on the core from the shell reactors, increasing the core mass. As we saw earlier, the mass of a star supported by electron pressure cannot exceed 1.4 solar masses. Once this critical mass is reached the electrons in the core start to combine with protons and form neutrons, thus removing the vital electron degeneracy pressure. The core now has no choice apparently but to collapse down to the density where neutron degeneracy pressure can be switched on as a last resort. Naturally the detailed behaviour of the star is immensely complex at this point, and several conceivable descriptions have been given by theorists. None the less there is a general consensus that the catastrophic collapse of the core to the size of a neutron star must represent the onset of the supernova explosion itself. Having reached this point we can now explore, tentatively, the pyrotechnics of the spectacular outburst itself.

9.4 The supernova explosion

When the core mass reaches the point at which electrons and protons start to combine to form neutrons a disastrous outcome is inevitable. The fusion of electrons and protons is accompanied by the release of the weird elementary particles termed neutrinos as well as the creation of neutrons. It is hard to describe a neutrino in everyday terms, because it is basically a chargeless, massless, bundle of energy. Its main property is its negligible interaction with ordinary matter. Nothing happens to a neutrino when it encounters an object as large as the Earth, for example: there is a very high probability that it will just zip straight through at the velocity of light. Immense numbers of neutrinos pass right through you every second. But the neutrinos can carry energy away from the interiors of stars very efficiently, under most circumstances, and precisely their reluctance to join up with other elementary particles means that they will not be delayed in their travels.

Once the neutrons start to form in the dense core of an evolved star, the actual collapse of the core proceeds rapidly. So rapid is this implosion that it shrinks from thousands to tens of kilometres in a mere one-quarter of a second. If the core is rotating, as indeed it must be, the rotation rate speeds up by thousands of times, as the radius shrinks, in order to conserve angular momentum. (The pirouetting skater again—remember?) The plunge to a much smaller radius also releases a tremendous amount of gravitational potential energy, which appears initially as heat. Immediately the temperature rises to a very high value and this leads to conditions in which copious quantities of neutrinos are made. In principle these neutrinos can fly away and reduce the energy of the core. But the central material is so dense at this stage that even the unstoppable neutrino finds the exit tough. Rather than streaming out unhindered, the neutrinos are forced into a more leisurely diffusion as they dodge around densely packed nuclear particles. One possibility is that these retarded neutrinos are able to exert a pressure, and this pressure wave (or shock wave) sweeps the material outside the core before it, thus ejecting the outer layers of the star in an explosion, to produce a gaseous nebula.

Another mechanism concerns the nuclear reactions that might take place immediately after the sudden implosion of the core. Up to the instant of implosion the core is encased in the dead ashes of the spherical nuclear reactors. After the almost instantaneous implosion this debris must plunge down a void until it crashes into the tiny neutron core. It is this infall that might energize a whole new series of run-away chain reactions, leading to a cosmic nuclear holocaust. As in the neutrino model, the net result is the sudden release of a great deal of heat energy deep inside the star, and this energy is available to throw off the supernova envelope. Calculations for a 10 solar mass supernova indicate that the shock wave bursts through the star in under a second. Within this time almost the entire envelope is on the way out, speeding to the depths of space at thousands of kilometres per second. The same studies have also suggested that the prompt emission of neutrinos lasts for one-twentieth of a second, so they leave the core in a sharp burst. This prompt emission of neutrinos suggests that supernovae could be detected close to the instant of the outburst. However, supernova searches that are sensitive to the prompt emission have not notched up a single success.

The early part of the explosion is not instantly apparent to distant observers because the outer layers of the exploding star are opaque to light. Only when the envelope has expanded and the density consequently dropped sufficiently to make the gas transparent will a supernova be visible. So the guest star of A.D. 1054 must have collapsed some time before it shone out so wondrously. How long before? This question can now be partially answered from data gained on a star that burst out in the spiral galaxy in M101 in 1970.

Spectroscopists monitored supernova 1970g, to give the outburst its clinical title, over a period of months, and were able to plot the star's radius as a function of time. After the implosion of the core, the outer envelope raced into interstellar space at 5,000 kilometres per second. At maximum brightness it extended 3×10^9 kilometres, which would allow it to encompass the orbit of Uranus. This radius, at which the envelope stops being opaque, and the supernova becomes visible, was reached after ten days. Within a month the envelope had coasted out to 2×10^{10} kilometres—bigger than our solar system—and its temperature had decreased to 6,000 K, about the same as the Sun. Can you imagine a star bigger than the solar system and as bright per unit area as the Sun? If you can then you have visualized the Crab explosion when it was four weeks old!

In the frenzied expansion the envelope eventually over-reaches itself and becomes quite transparent. At this stage it cools rapidly. The apparent size of the visible star shrinks, and the apparent brightness begins its steady fall. As we noted in Chapter 1, this gradual decline of the light curve was followed by astrologers for nearly two years in the case of the Crab Nebula outburst. In its subsequent evolution the expanding envelope transforms itself into the glowing nebula now seen as a supernova remnant.

9.5 Nuclear furnaces in expanding envelopes

Before finishing the description of the supernova outburst itself, here is a brief sketch of another speculative yet potentially important process; explosive nucleosynthesis. After the precipitous collapse of the core a shock wave travels out through the envelope. It compresses the partially processed waste from earlier nuclear phases, and can detonate a whole new series of nuclear reactions. The high temperatures (billions of degrees) and densities (10^{14} kilograms per cubic metre) that result from the compression of the gas as the shock wave moves through it, favour a host of reactions between atomic nuclei and particles. These reactions do not take place during the normal evolution of a star. They only happen at the very end.

One important process thought to occur has been called the

r-process by nuclear astrophysicists; the r signifies that this is a rapid succession of chain reactions. This series of events can occur when there is a high density of neutrons. Supernovae are one of the favoured cosmic sites for the r-process since there have been plenty of them.

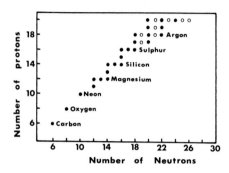

Fig. 35 Distribution of neutrons and protons in the low mass section of the periodic table. Filled circles are stable nuclei and open circles are unstable nuclei which decay to stable configurations by emitting electrons.

In the r-process atomic nuclei successively capture neutrons from a high ambient background of neutrons. Thus the seed nuclei, as they are termed, become progressively neutron-rich. When a neutron is added to a nucleus it may well become unstable. An unstable nucleus is liable to transform itself spontaneously into a different kind of stable nucleus. However, like all natural processes, such a spontaneous change takes time. The essence of the r-process is that neutrons are so numerous that the seed nuclei are likely to capture further neutrons before having time to reorganize themselves into more stable configurations. Eventually, of course, nuclei that are super-rich in neutrons result, and rapid decay to more stable forms becomes inevitable. The relaxation to a more stable state comes about through the emission of electrons; in effect some of the excess neutrons are turned into protons. Note that when the proton count in a nucleus increases (as it does when an electron is emitted), the nucleus moves up through the periodic table of the elements. Thus it is that the heavier elements in the periodic table are thought to be made in a time of less than one second in a supernova blast wave.

Under non-explosive conditions it is not possible to manufacture inside stars elements beyond the so-called iron peak. If you start with a star made of hydrogen and helium, then it will process these to create a series of nuclei up to and including iron and nickel. But it cannot synthesize the reactions needed to manufacture gold, silver, tin and lead (for example) because those reactions require an input of energy. Only when the physics gets out of control, in, say, a supernova outburst, are we likely to encounter the right conditions for the creation of heavier elements.

Astrophysicists currently consider that Type II supernova explosions (like the Crab explosion) are crucially important for the production of heavy elements in our Galaxy. The understanding of the processes is not uniformly good, but there is a general agreement between the predictions of the immensely complex computer programs and the composition of the real world of planets, stars and galaxies. This is a part of astronomy in which nuclear physics, mathematics, and observational astronomy converge to provide us with an outline of the history of matter. If the general ideas are correct then all the gold in the Bank of England and Fort Knox was originally made from base substances in great celestial explosions. The Crab Nebula is the remains of a cosmic alchemist's den.

9.6 Pollution of interstellar gas

The expanding envelope of a supernova rolls back the interstellar medium in its vicinity. Like a snowplough it pushes the material in front of it. Turbulence at the boundary gradually leads to a mixing of the interstellar gas and the supernova envelope. Eventually the expanding remnant sweeps up so much of the interstellar gas that any further expansion is essentially indistinguishable from the ordinary motions within interstellar gas. Then the matter that was once part of a star has returned to interstellar space, where the star formed in the first place. After a supernova explosion it probably takes tens or hundreds of thousands of years for the wreckage of a star to become an integral part of the interstellar medium once more.

The matter thrown off in a supernova eventually pollutes about one thousand times its own mass with heavy elements produced in nucleosynthesis. Viewing this on a cosmic time scale we can see that the interstellar medium gradually becomes enriched with the cast-off products of nuclear burning in massive stars. Indeed, it is perhaps wrong to speak of an undesirable pollution effect, since the enrichment is absolutely vital for the later formation of objects such as planets, plants and people. Bear in mind that low-mass stars like the Sun will never contribute significantly to the enrichment of the Galaxy. Although solar-type stars produce elements heavier than hydrogen, the reaction products remain forever trapped in the gravitational prison of a white dwarf star. As time goes on more of the elements in the Galaxy will end up in white dwarfs, where they are effectively cut off from any further possibility of participating in the formation of stars and galaxies. In the far far future much of the Galaxy will consist of white dwarfs and black holes.

The reaction products within interstellar clouds are presumably able to form dust and molecules, since we actually observe these in the denser clouds. It is very likely that the presence of dust and molecules assists the condensation of new stars from the gas. During the formation of a protostar, heat energy has to be transferred away from the contracting star; both dust and molecules can assist the efficiency of this cooling and thus aid the formation of new stars. Thus it is that the death of a massive star aids the birth of the new generation. For the Crab Nebula it will obviously be many many thousands of years before the heavy elements there get bound up inside a new protostar, but the probability is that this will happen eventually.

and it is bound to eventually, the chances are that the initial sighting will be made by an amateur astronomer who is searching for new comets or novae. Meanwhile professional workers have to manage with the meagre data from hundreds of years ago, supplemented by telescopic observations of remnants and supernovae in distant galaxies.

The search for supernovae in external galaxies is conducted photographically. The pioneer in this field, Fritz Zwicky, started in Pasadena, California in 1934. Much to the amusement of his colleagues he mounted an eight-centimetre camera lens on the roof of the physics laboratory at the California Institute of Technology. Then he proceeded to photograph the rich cluster of galaxies in Virgo at regular intervals. The principle behind this procedure was as follows. Even in the 1930s it was realized that exceptional stellar outbursts do occur rarely, and that these often exceed the luminosity of an entire galaxy for a few weeks. Therefore, with a regular photographic patrol of the galaxies, it ought to be possible to spot a supernova in another galaxy by comparing a series of photographs taken at intervals. Zwicky estimated that his equipment ought to be capable of picking up one supernova in a thousand years in a typical Virgo Galaxy. By looking at the great cluster in Virgo he hoped to capture two or three in the space of a couple of years. Unfortunately none were found between 1934 and 1936, when this particular experiment ceased. During the initial search Zwicky defined the procedures that have been used ever since for supernova patrols.

In 1935 an important instrumental development took place. Bernhard Schmidt of the Hamburg Observatory had invented a new astronomical camera with a wide field of view; we now call these instruments Schmidt cameras. Zwicky visited Hamburg, and on return persuaded his superiors that a Schmidt telescope ought to be constructed to aid the supernova search. Using part of a grant from the Rockefeller Foundation, George Ellery Hale (then the Director of the Mount Wilson Observatory) allocated $25,000 to the project. Within a year the redoubtable Zwicky had a 46-cm Schmidt telescope in operation. He took the first plate on 5 September 1936, and found his first extragalactic supernova in March 1937 in the spiral galaxy NGC 4157. An improbable event took place on 26 August 1937: with a dash of astronomer's luck Zwicky netted his second supernova, this time in IC 4182. The improbable factor was the intrinsic luminosity of the event: this supernova has not been equalled in brightness by any of the 400 objects subsequently logged, and indeed it outshone its parent galaxy by fully six magnitudes (a factor of 250).

Work with the 46-cm Schmidt continued and on average four extragalactic supernovae were discovered annually. In 1949 the powerful 1.2-metre (48-inch) Schmidt came into action, but it could not be used for supernova searches until 1959. This was because maximum priority for the use of the new telescope was assigned to the acquisition of the plates needed for the famous Palomar Sky Survey. However, once the larger Schmidt was applied to supernova searches the discovery rate shot up. As a result of Zwicky's work at Palomar, coupled with his efforts to get patrol work going on an international basis, over 400 extragalactic supernovae have been catalogued. Zwicky himself found over one hundred, an amazing achievement matching the efforts of the greatest observational astronomers. He speculated that this score probably entitled him to write a book on supernovae!

In the southern hemisphere the U.K. Schmidt telescope, which is similar to the Palomar instrument, captured its first extragalactic supernova in July 1977, purely by accident. Astronomers making a catalogue of unusual galaxies noticed a brilliant stellar image embedded in NGC 1411, which turned out to be a supernova. This telescope cannot start regular supernova patrols until it too has completed the important work of making a definitive photographic survey of the southern heavens.

Amateur astronomers could be organized to search for super-novae, just as they currently scan the heavens for novae. Supernovae in external galaxies are faint compared to the novae within the Milky Way, so that different search techniques would have to be used, and telescopes of moderate aperture employed. One proposal envisages a world-wide network of amateurs with telescopes of 25-cm aperture or larger. Although such telescopes are not exactly commonplace in the British Isles, where relatively high costs and poor weather daunt most amateurs, there are plenty of then in the U.S.A., Australia and Japan, for example. There are about 200 galaxies brighter than magnitude 12 that can be expected to have supernovae brighter than magnitude 12 at maximum. Most of the galaxies are spirals and an average supernova rate of one per thirty years is plausible. This sample of bright galaxies should therefore yield a handful of supernovae per year. The supernova searchers would be issued with a short list of nearby galaxies to be scanned on every possible night. Anyone thinking that they had found a supernova in one of 'their' galaxies would telephone or cable a central clearing agency to have the discovery confirmed and relayed to professional observatories. The advantage of using a large team of amateur astronomers, rather than relying solely on patrol photography, is that the opportunity to catch a supernova 'on the way up' is substantially increased.

It is by studying the extragalactic supernovae that a picture has been built up of the various types and their properties, and also an impression has been gained of how often supernovae occur in different types of galaxy.

Supernovae of Type I occur in both spiral and elliptical galaxies, and they are generally associated with the old Population II component of the stellar distribution. They are characterized by a mean absolute brightness of -19 magnitudes. The spectrum indicates expansion velocities of the order of 10,000 km per second and there are no conspicuous hydrogen lines. These supernovae are generally supposed to be due to the explosion of the highly evolved core of a star.

The Type II supernovae are more common. (Despite this, the first twelve supernovae found by Zwicky were all of Type I, a curiosity that illustrates how dangerously misleading are any conclusions based on small samples.) On average these supernovae shine out at -17 magnitudes, show hydrogen lines in their spectra, have a more moderate expansion of order 5,000 km per second, and are of higher mass than Type I. Type II supernovae are associated with the young Population I component that is found sprinkled along the spiral arms of the galaxies. These outbursts are triggered by the rapid evolution of massive stars.

In recent years much work on the statistics of supernovae has been performed by Gustav Tammann. A considerable amount is now understood about the types of galaxies within which supernovae occur frequently and so forth, but it is difficult to summarize this owing to the very wide range of properties encountered in the galaxies themselves. Thus, Tammann finds that the 'average' galaxy suffers an explosion every seventy-five years, but points out that there is no such thing as an average galaxy. A search sample of 408 bright galaxies is currently yielding an average of 2.5 new supernovae per year. On the other hand, the whole sky patrol programmes are discovering around eighteen fresh supernovae annually.

After making many adjustments it is found that supernovae are commonest in spiral galaxies with the arms widely spread out (type Sc in Hubble's classification scheme for galaxies). An example of such a galaxy is the Triangulum galaxy (or nebula), M33, which is a member of the local family of galaxies and is about two million light years from ourselves. Supernovae seem to be especially sparse in the tightly-wound spirals, type Sa.

Considerable controversy has surrounded the various attempts to estimate the rate of supernova explosions in our own Galaxy. A great difficulty arises for the following reason. The Sun is in the main plane of the Milky Way system, but well out to one edge. This means that almost all of the stars in the Galaxy are in fact concealed from our view by dense lanes of interstellar gas and dust. Typically it is only possible to penetrate for a few thousand light years through this haze, whereas it is a distance of one hundred thousand light years to the other side of the Galaxy. Hence we cannot easily scale from the observed rate of supernova explosions to get the true rate, although we can safely state that most supernovae located in the Milky Way are not visible from the Earth.

The six reasonably certain historical supernovae (1006, Crab, 1181, Tycho, Kepler and Cassiopeia A) yield a frequency of one every quarter century when applied to the entire Galaxy. Considerable uncertainty attaches to this estimate because of the small sample, but the value agrees precisely with the rate expected for a galaxy with the mass of the Milky Way system. Many attempts have been made to derive a rate from the observed distribution of supernova remnants. Recall that the ashes of a supernova consist partly of an energetic plasma, expanding into the interstellar medium, that becomes a strong source of radio waves. This radio emission is not absorbed by the gritty haze, so radio astronomers are able to see the effects of explosions that would have been visible on the Earth. Estimates from the population of supernova remnants range from about thirty-five to one hundred years per supernova.

Besides the glowing ashes, at least some supernovae result in a spinning pulsar as well. By looking at the statistics of the distribution of pulsars, astronomers have derived a birthrate for these celestial beacons of three per century at most.

For our Galaxy the various estimates point to a likely rate of one supernova every twenty-five to fifty years. The fact that none has been seen optically during a period ten times as long underlines the effect of obscuration and the feeling that one is well behind schedule. Note that our Galaxy comes quite close to the maximum rate, which is one every ten years for very luminous Sc galaxies. On the other hand the rate is down to one every million years in dwarf elliptical galaxies. Why are there these wide variations? The rate mainly reflects the different rates at which stars, especially massive stars, form. New stars are born out of the interstellar gas, which is abundant in Sc galaxies and non-existent in dwarf ellipticals. The star birthrate is much higher in the luminous spiral galaxies than the dwarf ellipticals. Hardly any stars are seen to explode in the latter; the massive stars that once inhabited the elliptical galaxies evolved and exploded long long ago. Now there are no young stars to replace them, and hence the stellar death rate (the mortality is far higher among young massive stars) has greatly declined. It is interesting to speculate whether we would have learned of the mysteries of supernovae if our Sun had been in a Galaxy with a much lower supernova frequency. The chances of a Crab Nebula nearby would have been extremely small.

10.2 The supernova of A.D. 1006

Oriental and Arabic sources record a guest star's appearance in the constellation Lupus on 1 May 1006, which may have been of greater apparent magnitude than any other supernova. Since this object was initially much brighter than Venus (possibly magnitude -9) and shone in the sky for several years it is a clear candidate for a supernova. In 1965 two Australian astronomers, F. F. Gardner and D. K. Milne pointed out that the shell-shaped radio source known as PKS 1459–41 coincides with the position of the star of A.D. 1006. The shell is typical of a supernova remnant, and indeed the interpretation of the event and the radio source as a supernova and its remnant are now regarded as certain. There is a reliable distance estimate of 1.3 kpc (about 4,000 light years), which makes this object the nearest of the historical supernovae to Earth. It is generally believed to have been a Type I explosion.

10.3 Tycho's 'Nova'

Aside from the Crab Nebula, the most famous of the historical supernovae is the new star watched by the great astronomer Tycho Brahe in 1572. He observed it regularly until it disappeared from view after 16 months. This new star was also carefully recorded in the official history of the Ming Dynasty. Tycho had the following to say about it in his book, *De Stella Nova*, published in 1573: 'Initially, the new star was brighter than any other fixed star, including Sirius and Vega. It was even brighter than Jupiter. . . It kept approximately the same brightness for almost the whole of November. On a clear day it could even be seen at noon.' The appearance of this supernova went strongly against the natural philosophy of the times, for it was widely held, in accordance with Aristotle's teaching, that the Heavens were perfect, and therefore unchanging. The sudden appearance of a bright star rather disturbed this simple philosophy.

Although astronomers searched for radio waves from the remnant, they did not find them until 1952. Two radio astronomers, R. Hanbury Brown and Cyril Hazard, who were then attached to the Jodrell Bank facility of the University of Manchester, detected a powerful radio source close to Tycho's position. After this discovery, in a series of pioneering observations on the identification of radio sources, Rudolph Minkowski found the optical remains using the great 5-metre Hale telescope on Mount Palomar.

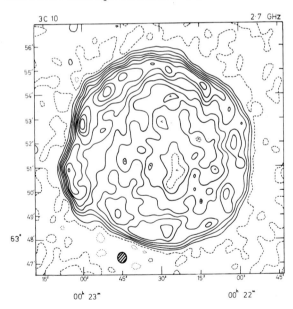

Fig. 36 Radio contours of Tycho's remnant mapped at 2.7 GHz with the Cambridge radio telescope.

As Fig 36 shows, the structure of the radio remnant is unusually symmetrical. Polarization maps indicate that the magnetic field spreads radially through this shell. Optical work has shown that the wisps associated with Tycho's remnant have been considerably decelerated in the interstellar medium. In fact the remnant consists of an expanding shell of relativistic electrons, which is responsible for the radio emission, surrounded by a thin and dense screen of hot gas that is being compressed by the expansion. This thin veneer of active plasma is responsible for the wisps that can be seen by optical telescopes. Finally, it is worth noting that Tycho was one of the first supernova remnants identified as an X-ray source, in 1967.

10.4 Kepler's 'Nova'

Hard on the heels of the 1572 guest star came the new star of 1604, called Kepler's nova. This did not manage to get as bright as Tycho's nova, but even so it became as brilliant as Jupiter, being visible by day for about one month and by night for a year. The historical records were accurate enough to enable Walter Baade of the Mount Wilson Observatory to discover the optical nebula in 1941. The optical nebula consists of bright filaments sprawled over about one minute of arc. Hanbury Brown and Hazard found the radio source associated with the supernova remnant. It is elliptical, with a thick shell, and has a diameter of three minutes of arc. Kepler's nova is definitely a Type I supernova, in fact, and it was located towards the galactic centre, at a distance of 9.5 kpc.

10.5 Cassiopeia A—an unseen supernova

A remarkable piece of scientific detective work has shown that the youngest galactic supernova remnant that we know about was apparently not noticed by astronomers. This is the supernova that occurred in Cassiopeia about three hundred years ago. Not until the new science of radio astronomy did astronomers have any inkling that a star had blown up in this constellation in historic times.

The pioneers of radio astronomy working at Cambridge in 1947 found the brightest source in the whole sky at metre wavelengths not far from the galactic equator and called it Cassiopeia A. Intensive efforts were made to pinpoint the object more precisely so that it could be matched to an optical counterpart. All efforts failed until a

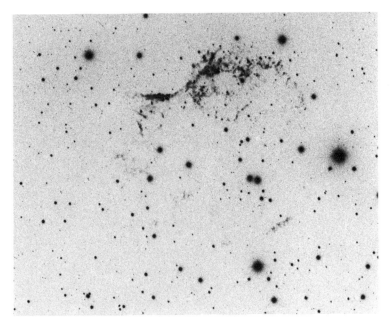

15. Cassiopeia A, the youngest supernova remnant in our Galaxy, photographed by
S. van den Bergh. (Courtesy S. van den Bergh and the Hale Observatories)

new radio interferometer came into use at Cambridge in 1950. Then
F. Graham Smith accurately measured the position of the Cassiopeia
object. This guided David Dewhirst of the Cambridge Observatories
to photograph the neighbouring region of sky and find a faint but
curious smudge of nebulosity. Dewhirst's findings were sufficient
incentive to spur Walter Baade and Rudolph Minkowski to
photograph the area with the giant Palomar 5-metre telesope. They
found a remarkable new type of emission nebula on their plates, a
disc of gas four minutes of arc across, and only plainly discernible on
its northern edge. Spectroscopy soon confirmed that they had
discovered an energetic supernova shell.

Within two years American astronomers had found that reddish
threads of gas in Cassiopeia more or less stood still, whereas blue-
green filaments moved outwards at a quite amazing angular speed.
Following the motion back in time suggested that the filaments
converged about A.D. 1667.

Canadian astronomer Sidney van den Bergh has made a special

study of the runaway knots of gas in the Cassiopeia remnant, using a series of photographs gathered during the first quarter century of investigation by optical astronomers. This has shown that fast knots grow and fade on a timescale of a few decades. For this reason it is sometimes difficult to follow the motion of a particular filament for more than a few years.

Work by van den Bergh and his collaborators indicates that the system of red knots of gas is in fact expanding quite slowly. Back-tracking its motion uniformly gives an origin of 11,000 years ago, which raises the quite intriguing possibility that expansion of a circumstellar shell may have preceded the final stellar cataclysm. By graphing the velocities of the high-speed filaments van den Bergh brackets this final disaster between 1653 and 1671.

An important spectroscopic investigation by Mexican researcher Manuel Peimbert gives abundances for the elements argon, oxygen and sulphur. These are an impressive forty times more plentiful in the Cassiopeia remnant (relative to hydrogen) than in the gas clouds such as the Orion Nebula that are the sites of star formation. This is observational proof that heavy elements really are manufactured and released as an integral part of the supernova process.

The Cassiopeia remnant is finding the going tough as it expands out. We know this because its radio emission is getting measurably weaker year by year. Expansion sapped the magnetic and thermal energy of the growing sphere of gas. Daily the radio waves get feebler. Radio maps display a beautiful shell structure.

Nobody saw the Cassiopeia explosion because it took place in a part of the Galaxy that is heavily obscured with dust from our viewpoint. The remnant shines through a layer of haze that dims light to almost one per cent of its initial intensity by the time it reaches the solar system. This drastic diminution appears to be the main reason why it was not recorded. Its distance is still highly uncertain, but is most likely around 8,000 light years. At this distance the implied velocities of the expanding filaments are in the region of 5,000 kilometres per second.

10.6 Remnants from the remote past

The remnants of the last thousand years have taught us what a supernova shell looks like. Many ancient shells, the supernovae of prehistory, are known as the result of radio and optical surveys.

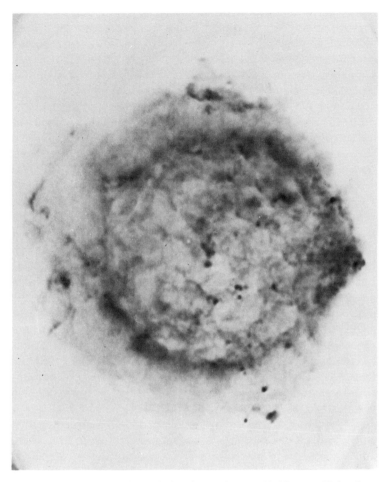

16. Radio map of Cassiopeia A, displayed as a 'photograph'. (Courtesy University of Cambridge, Mullard Radio Astronomy Observatory)

The most beautiful of all the ancient remnants is the Cygnus Loop, a cosmic question mark constructed from finest filagree. It is perhaps 20,000–100,000 years old and is nearly three degrees across. Radio maps show the familiar (obligatory!) shell-like architecture which correlates excellently with the optically-visible filaments. The Loop is only 2,000 light years from Earth; its progenitor explosion could have been quite a stunning sight to early man. The Loop is an X-ray source of some interest because of the excellent match between the structure at X-ray optical and radio frequencies. Its X-ray spectrum suggests that the high energy radiation is generated by heating as the gas of the remnant collides with the interstellar medium. At the centre of the Loop there is a source of X-rays that could be the compact relic of the exploded star, but it is not detectable as a pulsar.

Another old supernova remnant, Puppis A, is studied in the southern hemisphere. It is a radio and X-ray emitter with a more or less spherical appearance. Faint optical filaments are there all right, but not enough work has been done to get an expansion age. Other signs suggest an age of 5,000 years.

The title 'Crab Nebula of the Southern Skies' we must surely accord to the magnificent Vela supernova remnant. It doesn't look very like the Crab Nebula on optical photographs because it is much older; in fact it has fragmented during its expansion to give a broken shell. But in its own way it's just as nice to look at. The relationship between optical, radio, and X-ray filaments is less clear cut, perhaps because the remnant really is running out of energy and generally dispersing into interstellar space. Pulsar 0833–45 lies near to the centre of the delicate shells and is widely held to the compact star related to the supernova remnant. In a brilliantly executed experiment observers at the Anglo-Australian Telescope showed that this second-fastest pulsar is an optical flasher too. This demonstrated that the Crab Nebula pulsar is not entirely unique, which is perhaps reassuring for those who ponder the theory of such things: it's good to be able to get a second opinion on your models.

Vela is part of an enormous jumble of emission nebulosity that makes up the Gum Nebula. This was discovered by Colin S. Gum, an Australian, who published a paper on the nebular complex in 1955. He estimated it to cover a large area of sky, possibly 40 by 30 degrees, sprawling through the southern constellations Vela and Puppis. The outer dimensions may be as much as twice as big again. Tremendous amounts of energy are needed to ionize so large a volume of gas—perhaps 10^{45} joules. Some theorists feel that the Gum Nebula is a series of shock waves in the interstellar gas that were set going in one or more supernova explosions. Some of this gas is now kept excited by several very hot stars embedded in the nebula.

In this short sketch only a few remnants can be described. Many dozens are in fact available for close investigation in both hemispheres of the sky, and much work remains to be done,

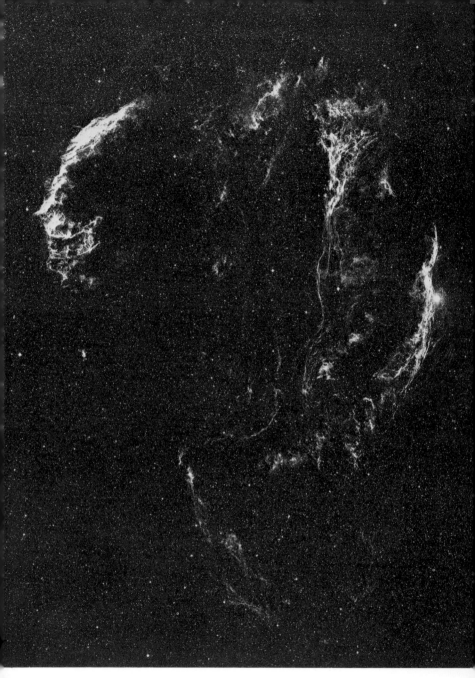

17. An ancient supernova remnant, the Cygnus Loop.
(Courtesy Royal Astronomical Society)

especially in regard to the optical spectroscopy of the remnants. Another challenging research area is the investigation of the supernova remnants in the Magellanic Clouds. These satellite galaxies to the Milky Way really only became available for close scrutiny in the mid-1970s, with the commissioning of several new telescopes in the southern hemisphere.

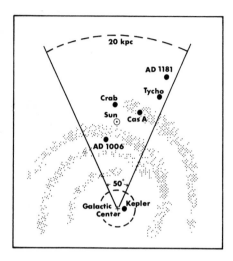

Fig. 37 Location of the main historical supernovae relative to the Sun, the spiral arms and galactic centre. Supernovae outside the 20 kiloparsec wedge presumably take place but cannot be seen on account of interstellar obscuration.

11 · The Crab Nebula and modern astronomy

11.1 First of the many

There is no doubt that the study of the Crab Nebula has made a bigger impact on the development of astronomy in recent times than the investigation of any other single object beyond the solar system. This nebula is truly a physics laboratory on a grand scale. In it we can see the four fundamental forces of physics played off against each other in eternal conflict. Weak and strong nuclear forces are at work in the neutron star, gravitational forces dominate in its vicinity, and the electromagnetic forces reign supreme in the magnetosphere and the nebula itself. What stroke of luck placed the Crab Nebula in the northern hemisphere, which until recently had a range of observational facilities far out-stripping those in southern countries? There is a remark attributed to Geoffrey Burbidge and often quoted thus: 'You can divide astronomy into two parts: the astronomy of the Crab Nebula and the astronomy of everything else.' We have looked at the astronomy of the Crab Nebula; now we shall see how it has influenced the astronomy of everything else.

The Crab Nebula is a bright object across a great sweep of the electro-magnetic spectrum. This accounts for the fact that the Crab Nebula was frequently among the first objects to be identified each time a new window on the universe has opened up.

The first suggested identification of a discrete radio source with an optical object came in 1947 from Australia. John Bolton and G. J. Stanley of Sydney suggested that radio source Taurus A could be matched to the Crab Nebula. Initially this finding did not cause any stir among astronomers in the northern hemisphere. They either felt that the positional information from far-off Australia was not good enough, or rejected the notion because it did not fit their own ideas as to what radio sources might be. Even four years after the initial suggestion, one of Britain's leading radio astronomers said that the coincidence of Taurus A with the Crab Nebula should not be taken too seriously. He went on to say that, although there might be some connection between the two, the identification could not be taken as evidence that radio sources in general were to be identified with stellar relics. This latter opinion we know with hindsight to be the correct standpoint as most of the radio sources are in fact associated with objects far beyond the Milky Way.

Later in 1951, however, David Dewhirst of the Cambridge Observatories was able to report to a meeting of the Royal Astronomical Society in London that the coincidence of Taurus A and the Crab Nebula was much more certain. This improvement came about because Graham Smith had used a radio interferometer at Cambridge to obtain more precise positions of several radio sources. In September 1951 Walter Baade of the Mount Wilson Observatory wrote to Graham Smith to relate the latest optical research on the optical objects that matched Smith's accurate positions. In the letter he stated that he had no doubts as to the correctness of the identification, and urged a radio search for emission from other recent novae. Smith replied that a sweep of the region where Kepler saw a supernova in 1604 produced a null result. Not long afterwards Walter Baade had more exciting news: he found that a radio source in Cassiopeia was close to an exceedingly interesting object, the intricate filamentary structure of which bore a strong resemblance to the Crab Nebula. It is now firmly established that the Cassiopeia source is, in fact, a supernova that went off a few hundred years ago. And so it was that the Crab Nebula pointed the way to the discovery of the youngest known supernova remnant in our Galaxy.

Fortunately a handful of other radio sources was identified in the early 1950s: Virgo A with the greatest elliptical galaxy M87, Cygnus A with a disturbed galaxy, and Centaurus A with a giant elliptical (in fact the first radio galaxy correctly identified as such). Had this not been the case one can but speculate that radio astronomers might have been led all over the Galaxy on a wild goose chase to fit their radio stars with supernova remnants. The realization that some external galaxies were strong sources actually had a much greater effect on the acceleration of radio astronomy than the Crab Nebula work. Once the distance to a few galaxies had been determined it was clear that the radio galaxies involved energy sources millions of times greater than any discrete object in the Milky Way, and so the emphasis moved out to the realms of the galaxies.

In 1956 the Crab assisted in the development of the new technique of lunar occultation. This phenomenon occurs when the Moon passes in front of a radio source. At any time the position of the Moon is known to great accuracy, so that the measurement of the radio signal and its variation with time as the occultation proceeds provides a method of finding positions and elementary information on the structure of a distant radio source. Experiments on the Crab Nebula by French and Australian groups demonstrated the feasibility of what became a powerful radio technique using relatively simple equipment.

In the visible part of the spectrum we can mark out the Crab Nebula as an interesting object from the very beginning of nebular spectroscopy. As mentioned in Chapter 3, it provided an easy target in difficult times and intrigued the early astrophysicists. Later, in the 1950s, it stimulated the use of polarization studies for the detection of light generated by the synchrotron process, and was in fact the first cosmic source of synchrotron light to be recognized as such.

By the time X-ray astronomy came on the scene the Crab Nebula was already firmly established as an important high-energy object. Then once again it provided an easy target, but by now there was a wealth of data at other wavelengths. So considerations that are not strictly scientific come into play. Space astronomy is expensive and to some extent competes for funding with ground-based astronomy. Therefore it is important when drawing up the case for a rocket and instrumentation to show in advance some of the scientific benefits. The Crab Nebula assisted these arguments during the birth of this new science, since it could be shown that X-ray studies would complement the volumes of data accumulated by spectroscopists and radio astronomers, as well as furnishing further information on the evolution of supernova remnants. Additionally the Crab Nebula turned out to be of intrinsic interest at the longer wavelengths, especially in respect of its shrinking size at shorter wavelengths. When the pulsar was found, that again contributed new impetus at a very opportune time when new X-ray, gamma-ray, and radio telescopes were coming into action. If the flow of funds is to be maintained in any big science, then a steady stream of new and interesting results is essential; the Crab Nebula is a significant tributary in this overall pattern of money flow.

11.2 The theorist's toy

There are not many astronomers, apart perhaps from a few cosmologists, who enjoy working on problems that are seemingly totally divorced from the real world of planets, stars and galaxies. For the healthy advance of all science, observation and theory must relate constructively, each pointing the way to new research fields, and each restraining the other. This interaction has taken place successfully. The Crab Nebula has provided valuable information and constraints on a wealth of theoretical activities related to supernova, spectroscopy, theory of synchrotron radiation, and neutron star physics.

The existence of this natural source of the radiation aided the early exploration by Soviet theorists into the theory of synchrotron radiation. Soviet scientists have, in fact, continued to be among the leading developers of this subject ever since. The theory is first mentioned in an astronomical context by the Swedish scientists H. Alfvén and N. Herlofson, who in 1950 speculated that radio stars (as most radio sources were then inappropriately termed) might be powered by such synchrotron mechanism. I. S. Shklovsky, in 1952, demonstrated that the background radio radiation from the Milky Way could be accounted for as a combination of thermal and synchrotron emission. Shklovsky then turned his attention to the Crab Nebula and suggested that the radio emission there might also be generated by the action of very fast electrons in the magnetic field of the nebula. His prediction of polarization was testable but the theory looked implausible at the time. Nevertheless Dombrovsky got the vital result confirming Shklovsky's theory. Then there was no holding back other astronomers.

Jan Oort and colleagues of Leiden took up the synchrotron idea in respect of the Crab Nebula and used it to relate optical and radio data. They presented their new results at a symposium of the International Astronomical Union in 1955. The general picture received further confirmation when optical measurements revealed that the jet in the disturbed galaxy M87 is also highly polarized.

The synchrotron mechanism was firmly established by the time of the 1958 Paris symposium on radio astronomy, and it had been extended to explain the radio emission from excited galaxies. Antony Hewish (awarded the Nobel Prize for Physics) reported thus on this meeting: 'The most immediate impression of the final session, devoted to theories of radio emission, was the complete supremacy of the synchrotron mechanism. Since its stimulus by the optical polarization measurements of the Crab Nebula, and more recently from radio studies of polarization at the Naval Research Laboratory, other mechanisms seem to have been almost entirely discarded.' His colleague Martin Ryle (another Nobel Laureate) reiterated this same point in a lecture delivered at the Royal Society of London: '(Synchrotron theory) has been shown capable of explaining the radiation from a number of different sources, including the Galaxy, the colliding galaxies NGC 5128 and NGC 1316, the Crab Nebula and NGC 4482 (M87). The latter two sources are of particular interest, since part of their optical emission has been shown to be plane polarized, a result which implies radiation of light by the same mechanism. There is, therefore, little doubt that the mechanism is responsible for the radio emission from these particular sources.'

Synchrotron theory subsequently proved to be immensely valuable in the interpretation of the emission from radio galaxies and quasars. The theory furnished a means of estimating energies and magnetic fields in these exotic objects, and proved that the energy requirements were very great indeed. By the time a reasonable amount of information on radio galaxies existed the rudiments of a theory were already there, courtesy of Shklovsky and the Crab Nebula.

11.3 The pulsar and neutron star theories

The finding of the pulsar in the nebula put flesh on the bones of the theory of neutron stars, and persuaded astrophysicists to drop pulsating white dwarfs as the explanation of pulsars. Two of the leading advocates of neutron star models, Tommy Gold and Franco Pacini, kindly supplied me with their recollections of the heady days of 1968. These reminiscences illustrate the influence of the Crab Nebula pulsar on the propagation of what came to be the accepted theory.

Franco Pacini had worked on the activity of neutron stars for his doctoral thesis at Rome University in 1964. Subsequently he kept up his interest, and knew that other researchers had suggested that neutron stars might have immensely strong magnetic fields. Eventually Pacini submitted a paper to the international science journal *Nature*. In it he postulated that rotating neutron stars could emit low frequency electromagnetic waves, and hence accelerate charged particles. As long ago as the 1950s, Sir Fred Hoyle had pointed out that the neutron stars might spin fast enough to fling charged particles into space. Pacini went on to suppose that this might account for the mysterious source of continuing energy in the Crab Nebula. The paper argued that a neutron star in the Crab would rotate at about one hundred times a second, and that energy could be supplied to the expanding supernova shell. All of this work was accomplished before the discovery of even one pulsar. Tommy Gold later described the paper as 'most prophetic' at a conference held in June 1969.

Scientists usually hear of great discoveries at conferences or in professional research journals. However, Pacini first learned of pulsars not in a journal but in a newspaper report while on vacation in New York State. Gold, Salpeter and Pacini discussed the new findings back at Cornell University, Ithaca, New York.

Gold was toying with the rotating neutron star model for pulsars. He tried to persuade Pacini, and other astronomers, to consider this theory. Within three months of the initial announcement from Cambridge he outlined the idea at a symposium arranged in New York. Gold requested a mere five minutes to present his short and speculative paper, but the official organizers felt the idea to be so outlandish that they would not permit it time in the conference programme! Nevertheless Professor Gold succeeded in saying a few sentences about it, and at about the same time submitted the paper to *Nature* for publication. This was published on 25 May 1968, only five days after reception at the *Nature* editorial office in London. Most journals do well if they get exciting news out in under three months.

All the basic ideas that now link neutron stars with pulsars—the rotation, the spin-down, strong magnetic fields, high rotation speeds, relativistic particle acceleration—are mentioned in this description of Gold's. However, when *Nature* published a souvenir volume of pulsar papers in late 1968, two of the principal observers, F. Graham Smith and Antony Hewish wrote introductions that make it clear that they were then still in favour of vibrating white dwarfs. Smith writes: '. . . The pulsating white dwarf seems at present to be the only reasonable contender still holding the field . . .' Hewish concludes: '. . . it is encouraging that white dwarf vibration still offers a plausible explanation of these mysterious and fascinating bodies.' So, the rotating neutron star certainly did not immediately find favour among pulsar observers as the correct explanation of the phenomena. In fact Gold had no conspicuous or vociferous supporters.

Although Pacini and Gold were both at Cornell University at the critical time when ideas began to gel, they worked independently. Pacini was mainly interested in the connection (if any) between neutron stars and supernova remnants, while Gold tackled the clock and emission mechanisms. Gold directed operations at the Arecibo observatory, and he urged the observers to search supernova sites for more pulsars. At the same time he continued to present his lighthouse model by means of lectures and stressed that supernova remnants with fast pulsars lurking inside ought to exist. His elation at the discovery of the Vela pulsar by the Australians was tinged with the disappointment that Arecibo missed the complete Crab discovery.

According to Gold he had virtually no supporters, indeed fierce opposition ruled, until the very day of the announcement of the fast pulsar in the Crab Nebula. This bombshell showed the white dwarf models to be hopelessly inadequate. When the slow-down had been discovered too, the rotating neutron star model became generally accepted. This must surely be one of the best examples in astronomy of a single observation succeeding in clearly discriminating two rival theories. The discovery set the stage for a mushrooming of activity in the theoretical study of neutron stars and their interiors. The event also, of course, confirmed Pacini's foresight in connecting neutron stars and their energy outflow with expanding supernova remnants.

11.4 The far universe

High-energy astrophysics is concerned with the most energetic phenomena encountered in our universe. Its scope embraces supernovae, close binary stars, X-ray sources, black holes, radio galaxies, and quasars. An outstanding problem of the 1960s and 1970s has been the so-called energy problem for radio galaxies and quasars; quite simply, these objects need such a large supply of energy in order to account for their luminosities that it is difficult to see how conventional energy sources (e.g. stars) could be involved. The Crab Nebula has played an important part in pointing to a solution of this energy problem.

Radio galaxies, discovered in the later 1940s and early 1950s, are usually disturbed galaxies which have two enormous clouds of radio emission symmetrically disposed on either side of the optical object. Early optical studies showed that radio galaxies are a long way from our Galaxy: one of the strongest, Cygnus A, is half a billion light years away. With a knowledge of the distance and apparent radio luminosity of a galaxy astronomers can calculate the intrinsic luminosity, or rate of emitting energy, at the object. Typically this is 10^{39}–10^{41} watts, and it is thought that the emission persists at this rate for 10^5–10^7 years. The radio emission itself is generated by the synchrotron mechanism, and this offers a method of calculating how much energy is needed just to produce the radiation observed today. The synchrotron radiation theory shows that the most efficient machine for generating radio emission is one in which the store of energy is more or less equally divided between electrons and the magnetic field. Although no one knows whether the radio sources have the sense to be arranged in an efficient way, the most conservative estimate of the energy results if we assume that the sources are so constructed. For the most energetic objects about 10^{53} joules of energy are needed just to explain what is presently observed.

This energy demand far exceeds that of more familiar objects. The luminosity of our Sun is 4×10^{26} watts, and its total emission during its expected lifetime of 10,000,000,000 years will be about 10^{44} joules. A mere 2×10^{17} watts of the Sun's energy falls on the Earth's surface. The total amount of man-made power is in the region of 4×10^{12} watts. Finally, consider how much energy the Sun would produce if it could be completely converted into energy: this energy is calculated by multiplying the Sun's mass by the square of the speed of light (in a consistent system of units of course), and it is 2×10^{47} joules.

From these comparisons you can now see the scale of the energy problem in radio galaxies. Their store of energy appears to be as great as almost a million Suns after total annihilation. If, more reasonably, we express the energy store in terms of 'conventional' cosmic energy, it is equivalent to the total output of a billion sunlike stars throughout the entire lifetime of those stars. It is most improbable, therefore, that the exotic clouds of electrons and magnetic fields have their origin in ordinary stars, because there is simply no way in which they could produce sufficient energy of the right form in a short enough time. It is even implausible to suppose that supernova explosions could power the radio galaxies, for rates of thousands of conventional supernovae per year, sustained over thousands of years would be required, and even then most of this energy would have to emerge as high-velocity electrons.

With this discovery of quasars (or quasi-stellar objects) in 1963 this problem became, if anything, more acute. According to the majority of astronomers who have studied quasars, they are very remote indeed, perhaps billions of light years from Earth. A quasar is an object which has a starlike appearance but which is located at a very great distance—in the realms of the furthest galaxies. Most quasars are variable in their optical and radio emission, some on timescales as short as a few days. It is impossible for a source of radiation to switch itself on and off in a time shorter than the interval required for light to cross the source, so the primary energy-producing region in a quasar is only light days in diameter. In other words it is hardly larger than our solar system. The energy problem is therefore even more tightly constrained. The pyrotechnic display has to be located in a volume that is cosmically minute.

But in the Crab Nebula we have an observable example of a small region of space, only tens of kilometres across, out of which considerable energy—10^{31} watts—is emerging. This is puny compared to the demands of a radio galaxy or quasar, but significantly it is energy in the appropriate form of highly-relativistic electrons and magnetic fields. Although theorists have not elucidated every step that takes energy from the spinning neutron star and converts it into particle energy, the fact of the matter is that the very existence of the Crab Nebula showed high-energy astrophysicists that such a process does actually occur in the real cosmos.

A particular feature of pulsars that is relevant to quasar and radio galaxy theories is the magnetosphere. Around the twirling neutron star there is this vast world of plasma and fields. Here the energy transfer takes place. Energy that was once imprisoned in the rotation of the neutron star is mysteriously converted in the magnetosphere to relativistic electrons. The conversion produces the same type of energy distribution in both pulsars and quasars.

The discovery of the fast pulsar greatly stimulated theories of the extragalactic radio sources that attribute the ultimate energy source to gravitational collapse. Imagine the Crab Nebula scaled up to galactic dimensions. Can we conceive of the central regions of a galaxy (or more likely a protogalaxy) collapsing under gravity? Such a process releases a large amount of gravitational potential energy, which might then be dissipated through rotation, as has happened to the Crab Nebular pulsar, to re-appear as fast particles and fields. One structural feature strongly supports this concept. The twin radio clouds could match the poles of a massive rotating object, and thus explain the characteristic alignment cloud-galaxy-cloud. Although many of the deep questions connected with radio galaxies and quasars must still be regarded as wide open, the lessons learned from the Crab Nebula are extremely interesting and encouraging. Most of the events known to occur in the excited galaxies can be observed in miniature in the Crab Nebula.

Glossary of Terms

absolute magnitude The magnitude of a celestial body placed at a standard distance of 10 parsec.

absorption A decrease in the intensity of radiation as it travels through a medium.

absorption-line spectrum A spectrum that contains dark lines due to absorption.

abundance The relative proportion of a particular element.

ångström Unit used for measuring wavelength equal to 10^{-10} metres.

apparent magnitude Measure of the observed brightness of a celestial body as seen from Earth. This quantity depends upon the intrinsic energy flow from the body, the distance of the body from the observer, and the amount of any absorbing matter.

atom The smallest portion of a chemical element that can participate in a chemical reaction.

comet A diffuse mass of gas and solid particles in orbit around the Sun.

continuum (or continuous) spectrum A spectrum that contains neither emission nor absorption lines.

cosmic ray High-energy charged particle from outer space.

electron A fundamental particle carrying a negative charge which is a basic constituent of all atoms. The electrons form a cloud around the positively charged nucleus.

electronic transition A rearrangement of the outer electrons of an atom or molecule resulting in the emission or absorption of energy.

electron volt Unit of energy used in atomic physics, equal to 1.6×10^{-19} joules.

emission-line spectrum A spectrum that contains a series of emission or bright lines due to emission by atoms.

Faraday rotation Rotation of the plane of polarization of electromagnetic waves when they pass through electrons in the vicinity of a magnetic field.

guest star Chinese term for a comet, nova or supernova that becomes unexpectedly visible.

image tube An electronic camera in which electrons released by photons hitting a sensitive material, rather than the photons themselves, are used to take a photograph. This process is more efficient than traditional photography.

interference The cancellation or reinforcement of wavetrains from the same source of electromagnetic radiation.

interferometer An instrument for determining the angular structure of small sources of radiation by means of the principle of interference.

interstellar medium The gas and dust between the stars.

ionization Removal of electrons from atoms to make positively-charged atoms, or ions.

lunar occultation A lunar occultation occurs when the disc of the Moon cuts through the line of sight to another body.

magnetosphere The magnetized region of space above the Earth's ionosphere.

magnitude Arbitrary number used to indicate the brightness of an object. The scale is logarithmic. Two stars differing by five magnitudes differ in luminosity by 100. The larger the numerical value attached to the magnitude, the fainter the object in question.

moment of inertia Measure of the tendency of an object to resist a rotational force.

nanometre (nm) One thousand-millionth of a metre, used as a unit of wavelength.

nebula Term formerly used to describe all hazy patches in the sky, many of which are now known to be galaxies, star clusters and gaseous clouds.

neutrino A chargeless, massless, elementary particle that travels at the speed of light.

neutron An uncharged nuclear particle with a mass slightly greater than that of the proton.

neutron star An ultra-dense star composed primarily of neutrons.

nova A star exhibiting a sudden outburst of energy, typically increasing its luminosity by about 10–12 magnitudes, but exceptionally by as much as 20 magnitudes. Novae generally recover after the outburst, unlike supernovae.

nucleon Massive particle, generally a proton or neutron, found in an atomic nucleus.

nucleosynthesis Formation of an atomic nuclei in the nuclear reactors inside stars.

nucleus The central massive part of an object, e.g. atom, comet, or galaxy.

parsec Basic astronomical unit of distance. The distance at which one astronomical unit subtends an angle of 1 arc second. $1pc = 3.26$ light years $= 3.086 \times 10^{13}$km.

planetary nebula Shell of gas excited and made luminous by a central hot star.

plasma A gas composed entirely of ionized atoms.

polarization Electromagnetic radiation is said to be polarized when the electric and magnetic waves are caused to vibrate in a preferred plane. If the plane is rotating, then the polarization is elliptical or circular.

population I General class of young stars found in the disc of a galaxy.

population II General class of old stars found in galactic nuclei and globular clusters.

positron Elementary particle like the electron but carrying a positive charge.

power law A formula of the form
$$S(v) = R v^{\alpha}$$
in which the value of a quantity S measured as a function of the variable v, i.e. the value of $S(v)$, is proportional to v raised to a power which in this case is denoted by α. Such a law gives a straight line when plotted on logarithmic graph paper.

proper motion Motion of an object across the line of sight on the sky.

proton Massive elementary particle carrying a positive charge. The nucleus of a hydrogen atom is a proton.

proton-proton chain Series of reactions inside a star resulting in the change of hydrogen to helium with consequential release of energy.

pulsar Rapidly-rotating neutron star which emits a regular pulsed radio signal.

quantum mechanics Branch of physics dealing with the microscopic world.

quasar Compact starlike object with a high redshift.

quasi-stellar radio source Radio emitting quasar.

radial velocity Velocity measured along the line of sight toward the observer. The radial velocity is negative when the motion is toward the observer and positive when it is away.

relativistic Travelling at close to the speed of light.

shock wave Zone in which sharp changes occur in the physical properties of a fluid, often when the velocity of the fluid begins to exceed the speed of sound.

supernova When a massive star reaches the end of its life it cannot settle down without first going through a gigantic stellar explosion in which the star's luminosity may become as great as that of an entire galaxy for a brief period of time.

supernova remnant The gaseous nebula left behind after a supernova explosion. These are strong sources of radio emission.

synchrotron radiation The radiation emitted by relativistic electrons when they encounter magnetic fields.

transition Rearrangement of the outer electrons in an atom or molecule.

zodiacal light Faint pillar of light seen in the zodiac after sunset and before sunrise (also called the false dawn) which is due to the scattering of sunlight from particles in interplanetary space.

Bibliography of sources

While researching this book I used a superb bibliography privately assembled by J. R. Shakeshaft, to whom I am greatly indebted as it saved me much tedious work. About 300 research papers have been consulted as well as dozens of books. The full bibliography would take many pages of this book and is only appropriate to a research monograph. The books and papers listed here are either of special importance or will provide a good entry point to the daunting mass of literature that now exists.

In this bibliography titles of papers have been re-written more concisely.

Abbreviations

Ap.J. Astrophysical Journal.
M.N. Monthly Notices of the Royal Astronomical Society.

Baade, W. Optical observations. *Ap.J.* 96, 188, 1942. *Astronomical Journal*, 59, 355, 1954.

Bell, S. J. and Hewish, A. Angular size and flux density of small source at 81.5 MHz. *Nature*, 213, 1214, 1967.

Bok, B. J. Gum Nebula. *Sky and Telescope*, August 1971.

Brandt, J. C. *et al. Rock art records in Archaeoastronomy in Pre-Columbian America*, edited by A. Aveni, Texas Univ. Press, 1975.

Burn, B. J. Synchrotron model for continuum spectrum. *M.N.* 165, 421, 1973.

Chevalier, R. A. and Gull, T. R. Outer structure. *Ap.J.* 200, 399, 1975.

Clarke, D. H. and Stephenson, F. R. *The Historical Supernovae.* (This magnificant monograph gives detailed references to all the research on the oriental observations. This material is not listed separately in this bibliography.) Pergamon Press, 1977.

Davidson, K. Chemical abundances. *Ap.J.* 186, 223, 1973.

Davidson, K. and Humphreys, R. M. Comment on some spectrograms. *Publications of the Astronomical Society of the Pacific*, 88, 312, 1976.

Davies, R. D. and Smith, F. G. (eds) *The Crab Nebula: IAU Symposium 46.* (This large proceedings volume contains a wealth of interesting papers that are not listed separately in this bibliography.) D. Reidel (Dordrecht—Holland), 1971.

Duin, R. M. and van der Laan, H. Radio polarization at 21 cm. *Astrophysical Letters*, 12, 177, 1972.

Gorenstein, P. and Tucker, W. Supernova remnants. *Scientific American*, June 1971.

Hansen, C. J. (ed.) *Physics of Dense Matter: IAU Symposium 53.* (Many papers on neutron star matter.) D. Reidel (Dordrecht—Holland), 1974.

Hewish, A. and Okoye, S. E. Fine structure at 38 MHz. *Nature*, 203, 171, 1964. 207, 59, 1965.

Lampland, C. O. Changes in structure. *Publications of the Astronomical Society of the Pacific*, 33, 79, 1921.

Mitton, S. *The Cambridge Encyclopaedia of Astronomy.* (This book furnishes a wealth of background information on modern astronomy. The sections on stellar evolution and dense states of cosmic matter are of particular relevance to the Crab Nebula.) Jonathan Cape (London) and Crown (New York), 1977.

Morton, D. C. Neutron stars as X-ray sources. *Nature*, 201, 1308, 1964.

Murdin, P. and L. *The New Astronomy.* (Despite its title this is a popular account of supernovae.) Reference International (London) 1978.

Nature reprint volumes. *Pulsating stars*, vols. 1 and 2. Published by Macmillan (London) in 1968 and 1969. They contain many of the early papers on pulsars.

Oort, J. H. and Walraven, Th. Optical polarization. *Bulletin of the Astronomical Institutes of the Netherlands*, 12, 285, 1956.

Rees, M. J. and Gunn, J. E. Origin of magnetic field and relativistic particles. *M.N.* 167, 1, 1974.

Scargle, J. D. Centre of activity. *Astrophysical Letters,* 3, 73, 1969.

Scargle, J. D. Activity at centre. *Ap.J.* 156, 401, 1969.

Scargle, J. D. and Harlan, E. A. Activity following pulsar spin-up. *Ap.J.* 159, L.143, 1970.

Shklovsky, I. S. *Supernovae.* (A classic book with much material of historical interest. Hilariously translated into English.) John Wiley and Sons Ltd., 1968.

Trimble, V. Motions and structure of filaments. *Astronomical Journal,* 73, 535, 1968.

Trimble, V. Ionization and excitation. *Astronomical Journal,* 75, 926, 1970.

Trimble, V. and Woltjer, L. Mass. *Ap.J.* 163, L97, 1971.

van den Bergh, S. Jet-like structure. *Ap.J.* 160, L27, 1970.

Wilson, A. S. Radio structure. *M.N.* 157, 229, 1972. *M.N.* 160, 355, 1972. *M.N.* 160, 373, 1972.

Wolff, R. S. *et al.* Spatial structure of X-ray source. *Ap.J.* 202, L15 and L21, 1975.

Woltjer, L. Optical spectroscopy. *Bulletin of the Astronomical Institutes of the Netherlands,* 14, 39, 1958.

Index